Praise for Soul Evolution with Zarathustra: Wisdom for 2012

"Soul Evolution with Zarathustra brings the unseen into focus, and the unknown into awareness. Carolyn Cobelo is an unusually gifted seeker of wisdom and truth."
- **Max Highstein**, author of *Intuiton Retreat* and bestselling new-age music composer.
www.maxhighstein.com

"The step-by-step exercises offer guidance and an ease of movement along the Ascension Path. A must-read guide for human beings transforming through the intense energies leading to mastery in 2012 and beyond."
- **Cathy Rosek**, CSC, HTP, author of *"Who Am I and Why Am I Here"*

"A masterpiece. The exercises are exquisitely helpful for so many issues people are facing today."
- **Dr. Deborah Sie**, NMD, award-winning naturopathic physician.

Also by Carolyn Cobelo

BOOKS
The Power of Sacred Space:
Exploring Ancient Ceremonial Sites

The Spring of Hope: Messages from Mary

AUDIOBOOKS:
Crossing Over: A Journey to the Light
Angels: Connecting to Your Guides in Spirit
Creating the Life You Want

eBOOKS:
The 2012 Companion

SOUL EVOLUTION WITH ZARATHUSTRA

Wisdom for 2012

Carolyn E. Cobelo

with

The Spirit of Zarathustra

Copyright 2011 by Carolyn E. Cobelo
All rights reserved.

Published by Akasha Entertainment
Carmel, CA

Edited by Jayn Stewart
Art Direction by Raul Chico Goler

ISBN-13: 978-1453688557

www.AkashaEntertainment.com

DEDICATION

This book is dedicated to the god of creative dynamic force, Ahura Mazda, the All Mighty Holy One.

The book's purpose is to share the wisdom that I have received through my relationship with my Spirit guide, Zarathustra, and to inspire others to seek their own spiritual guidance. It is my hope that as you read these words you will gain fortitude and faith in your search for truth and your own wisdom. I invite you to join me in an adventure into your own Self.

The writing is presented in co-creation with the spirit of Zarathustra. Neither he nor I could have written this alone, for we each are in need of the other as transmitter. I do not feel that I am communicating with the personage of Zarathustra, but rather, vibrating at the same frequency as his energetic field. This resonance enables me to transfer the knowledge from him into a verbal concrete form.

It is recommended that you record the exercises in the book for later playback. You may that find reading these words out loud enhances their vibratory effect.

With blessings and with love, I welcome you.

Carolyn E. Cobelo C.S.W.
Director of Akasha Entertainment, LLC
www.AkashaEntertainment.com

TABLE OF CONTENTS

1. The Soul's Purpose	1
2. The Seven Layers of the Aura	5
3. The Seven Chakras	7
4. How to Channel	15
5. Spiritual Guides	17
6. The Soul	21
7. Archetypal Identification	23
8. Archetypes and Spiritual Guides	27
9. Alchemy	29
10. Alchemy of Sexuality	33
11. Projection	41
12. Transference	45
13. Transcendence	49
14. Transformation	53
15. Transmutation	57
16. Ten Steps to Transmutation	61
17. Separation	67
18. Pain	71
19. Transitions	77
20. Birth and Akashic Records	85
21. Life Energy	87
22. Power	89
23. Approaching 2012	91
24. Addictions	95
25. Dreams	101
26. Responsibility and Commitment	105
27. The Wave of 2012	109

CHAPTER ONE
THE SOUL'S PURPOSE

Within each human being is a seed of potential from which the soul's purpose emerges. In order for this purpose to be released, there must be the release of the obstruction of energy on four basic levels: physical, emotional, mental, and spiritual.

The soul's purpose is contained within the energy field of each human being as a seed, or blueprint, which can expand and grow to fruition as appropriate circumstances are set in place. These circumstances involve the release of fear to allow the emergence of the unique combination of characteristics and abilities that are designed to create a contribution to the evolution of the Earth—and, in the larger sense, a contribution to All That Is. This purpose is carried through lifetimes, with each lifetime offering the opportunity to clear obstacles to full expression of the soul.

Releasing fear helps release love and creativity, which, in turn, allow the soul's purpose to come into material form. Fear restricts us by limiting physical action, emotional expression and spiritual awareness. Fear develops as a response to anticipated physical or psychic danger. It is recorded in memories that influence the interpretation of current events. These memories may be of this lifetime (conscious or unconscious) or past lifetimes.

When these responses to danger become inverted, negative self-images and beliefs form. These images and beliefs set up negative behavior patterns. Often referred to as the Shadow, they are revealed in dreams, fantasies, guided imagery, and journal writing. Fear blocks their emergence, but as the Shadow breaks through into the Light of consciousness, the fear dissipates, thus releasing creativity and love.

Physical

We experience fear on physical, emotional, mental and spiritual levels. On the physical level, symptoms of fear include muscle contractions, anxiety and rapid heart beat, perspiration, constriction of muscles in the extremities, dry mouth, spastic colon, constipation, diarrhea, nausea, tightness in the chest, and faintness. Fear causes the body to prepare to fight or flee in the face of anticipated danger. These fearful reactions can become chronic, creating the belief that this is a normal state. As adjustment to the conscious state of fear proceeds, consciousness of fear moves into the unconscious, where it becomes hidden from everyday thought. Thus, the human being is in a constant state of fear without being aware of it, but the body knows it and reacts with other physical symptoms.

Emotional

On the emotional level, fear expresses itself in depression, anxiety, or both simultaneously. Emotional symptoms include crying fits, panic attacks, lack of (or excessive) interest in sex, lack of interest in food and/or exercise, lethargy, nervousness, sleep interruptions, and jitteriness. These symptoms are difficult to ignore, and for many people, they seem to be just part of life.

Mental

Fear manifests mentally as negative thought patterns that are generally critical of oneself or others. These thoughts are concerned primarily with protecting us from anticipated danger. Examples of negative thought patterns are: "I'm not good enough," "I will fail," "No one really likes me," "I am unlovable," "Nothing good will happen to me," "I am a victim." The list goes on and on. These negative thought patterns limit us in taking risks, activating creative energy, expressing ourselves freely, and attracting love.

Spiritual

On the spiritual level, fear creates distrust of the divine benevolence of God. This is often translated as a sense of being betrayed by God. The fear of separation from the whole is the foundation from which all other fears emerge. This fear of separation is basic to all human beings in that we are all incarnated from a state of unity into a separate physical body. This creates detachment from the whole and from other bodies. This fear prevents the bliss of union with the Divine.

The desire to know the soul's purpose often intensifies during mid-life. This desire does not usually become conscious immediately. We may sense it as vague anxiety, depression, dissatisfaction with the status quo, and attempts to hold onto youth and the past. At this time the natural flow of energy turns inward. If this flow is resisted, the symptoms discussed above may arise. As we release the fear of entering into this flow of energy, we begin to discover the glory of our souls and the purpose for which our souls have incarnated.

The word *Akasha* refers to the records of All That Is and Ever Will Be in the form of translucent particles of light. Knowledge of these records is accessible through high levels of consciousness. Here soul energy vibrates with other soul energies to create movement toward Divine Creation. This is where the soul's purpose is created, maintained, and dissolved in an evolving process of Divine Manifestation.

The chasm between that which we believe to be true and that which is true is related to our knowledge of ourselves. We may identify with the physical self and believe that the material reality is the only truth, or we may identify with the All Mighty One and believe this to be the truth. Most of us are between these two realities.

As investigation of the human energy field expands, understanding of the complex reality of human existence alters to meet these findings. Since measurement of the human energy field other than on the physical level is scarce, science knows little about the realities beyond the material. This is developing, however, and will at some point correspond with the truth as it is known from the spiritual dimension. Ultimately, no measurement is possible, for the measurer would be included in the measurement, and there would be no difference of form or space. This is the true reality.

Message:

We hail the wisdom of the ages and this meeting of the light of spirit. Time moves to en-wisdom us. So, let the spaces of form move toward the light. Hail forth, Heavenly Father. Move earth and stone to show the way. Know that the light you see is but a reflection in denser form of All That Is and know it is for the mind to comprehend that these definitions issue forth. In truth, all light forms reflect the One in multitudinous delight

CHAPTER TWO
THE SEVEN LAYERS OF THE AURA

The layers of the aura, which is the light body surrounding and interpenetrating the physical body, are seated in the chakras. These layers are bands of color that surround and interpenetrate the physical body. The chakras are localized in different regions within the same space as the spinal column. The layers of the aura and the chakras are two interrelated aspects of the human energy field.

First and Second Layers

The first layer is the physical body. It is made up of dense organic mass held together by the second etheric layer, which exists as a net of brilliant threads of light in the shape of the physical body. Within this layer pulsates the life energy of electric pulse, chemicals, and flesh, compacted into form. The second layer, the etheric body, encompasses and interpenetrates this mass, holding sensations of pleasure and pain. It pulsates in multi-colors, expressing and holding all emotional experience of this incarnation, including dreams.

Third Layer

The third body consists of a layer of thoughts. This is where thought forms ebb and flow, attracting like thought forms to them. These thoughts form beliefs which then govern action. As we develop spiritually, positive thoughts increase. Negative thoughts decrease or are repelled if they are sent by someone else.

Fourth Layer

The fourth body holds the force of love and truth and connects to the higher realms. It is known as the astral body. This is the spiritual world of light beings such as cherubs, spirits of the natural environment, guides, angels, and others of different dimensions. This world is sometimes known as the "land of summer" which exists as the pastel landscape of the causal plane.

Fifth Layer

The fifth body is associated with sound and creative expression. Here is where we hear the voice of angels and guides and where we witness spiritual operations taking place. The "doctors" are administrators to the physical body in a higher dimension, which we can see in color and movement. The operations resemble surgery as we know it.

Sixth Layer

The sixth body holds visions of worlds unseen by the human eye. They are brought to consciousness by the Third Eye in the forehead. Here spirits take form in light and love from angelic sparkles as the nectar of God. Unconditional love flows in knowledge of the reality of the All Mighty One.

Seventh Layer

The seventh body is the home of our spiritual guides. It is a meeting ground for us and the Divine Mind or Divine Consciousness. Here is where the Self joins with the Divine in mystical union.

CHAPTER THREE
THE SEVEN CHAKRAS

In order to live we must absorb energy from the greater atmosphere. Continual fueling from the vital force within the atmosphere provides the energy for cellular metabolism. In order to function in material form, we require constant metabolism. The energy for this metabolism is absorbed into our systems through vortexes of spinning energy, referred to as chakras. These major chakras are located in a line from the base of the spine to above the crown of the head. Each vibrates at a different rate. Each spins like a wheel in relation to adjacent chakras.

Each chakra provides a different vibration of metabolism, which reflects different physical, emotional and spiritual influences. The chakras appear different if perceived from different levels of vibration. There are many more than ten chakras, and the perception depends on the perceiver. Here we will consider only the first seven major chakras.

The First Chakra

The first chakra is located at the base of the spine. It holds the basic life force, the will to live, and basic sexual and power drives. Its color is red, and it is related to the adrenal glands, spinal column, and kidneys. This is the seat of the etheric body, which forms a web of shimmering gray-blue lines of light, upon which the physical cells of the body grow. It records the memory of birth trauma and life-threatening traumas in childhood or later in life. These traumas may include child abuse or neglect, experiencing or witnessing a serious accident, and living or fighting in a war-torn environment. This chakra relates to the element of earth and the sense of smell. When this chakra is open and flowing well, one feels confidence and pleasure in the ability to survive, with a strong presence of being.

Second Chakra

This chakra sits at the level of the gonads. It is the seat of the emotional body and dreams. The second chakra holds our basic emotions including passion, rage, frustration, and sexuality. It relates to family, groups, the collective, and personal creativity. This chakra vibrates in the color of orange and red-orange. It is related to the element water and the sense of taste. The areas associated with this chakra are the reproductive system and lower back. When there is restriction in the second chakra, one may experience pain or disorder in the reproductive system and/or lower back. It may manifest emotionally as lack of sexual interest or excessive need to control others through sexuality. Mentally, there may be either preoccupation with sexual thoughts or lack of these thoughts. We may have negative or conflicting thoughts about our families or the groups of which we are a member. Opening this chakra brings sexual excitement and pleasure and openness to life and sexual love

Third Chakra

The third chakra is located in the solar plexus and is the seat of the mental body. It is associated with the linear mind, the ability to analyze and digest information, will power, and responsibility. Some say that New York City is the third chakra center of the world. This chakra is the color of yellow and is related to the pancreas, stomach, gall bladder, liver and nervous system. Pain or dysfunction in any of these organs suggests energy blockage in the third chakra. It is associated with finding one's place in the universe and the use of power and control. The opening of this chakra brings a sense of expansiveness and confidence in the ability to meet life's challenges. It brings joy and delight in the merging of rational, logical thinking and spirituality.

We can maintain control easily and let go of control when we wish.

Fourth Chakra

The fourth chakra resides in the center of the chest and is associated with the heart and the astral plane. This is where physical and spiritual realities join. Energies related to love and love of the All Mighty One dwell here. When this chakra opens, we vibrate to meet the One and easily experience the impact of change and transition. This chakra is related to the element of air, the colors green and pink, and the sense of touch. Over-dependency on the heart chakra brings over-identification with the suffering and struggles of others. It is connected to the thymus gland, the heart, the circulatory system and the blood. Pain or dysfunction in any of these organs suggests a block in the energy of this chakra. Fears of intimacy, love, and betrayal block this chakra. When the heart is open, we feel love for the Self and for humanity, compassion, personal power, and safety as we rest in the arms of the All Mighty One. It is through our heart chakra that we connect ultimately with the All Mighty One.

Fifth Chakra

The fifth chakra dwells in the throat. It is associated with sound, communication, creativity, professional will, speech, and the sense of hearing. The ear is literally a channel for vibrations and is the only totally receptive sense organ. All the other organs of sense are expressive and receptive. In this chakra we hold the ability to assume and delegate authority and to overview planning without being consumed in the details. This chakra's color is sky blue. It is related to the thyroid gland, lungs, bronchial tubes, vocal apparatus, and the alimentary canal. It is associated with taking in and assimilating nourishment and love. Here we experience ourselves in society and in our profession, and when we hear the voices of our spiritual guides. We are able to sound, or tone, into the chakras for healing. Blocks in this chakra may be revealed by pain or dysfunction in any of the associated organs or in speech difficulties, professional inertia, fears of intimacy, and suffocation. When this chakra is open, we have confidence and a sense of well-being in our profession. We express ourselves with ease and delight. Chanting and singing assist us in opening this chakra. Channeling spiritual guidance and/or clairaudience occur when we release blocks in the fifth chakra.

Sixth Chakra

The sixth chakra, known as the Third Eye, is the intuitive psychic center and is located in the center of the forehead. Unconditional love flows from here. There is a sense of knowing and of perceiving Truth. Internal viewing, or x-ray vision, enables us to see into solid mass, such as the body and other material forms. This chakra is indigo blue or blue-purple light and is connected to the pituitary gland, the lower brain, left eye, nose, ears, and nervous system. When this chakra is open we have a sense of being able to separate from the world while penetrating the essential meaning of what is taking place—a state of being in the world but not of the world. From here spiritual visions emerge and we have the ability to manifest our wishes. Some headaches, eye strain, some mental illnesses, dysfunction in the pituitary gland or in the kidneys indicate blocks in the sixth chakra. Some people overuse this chakra because they have learned to rely on their intuition as a defense in childhood against abuse or neglect. This leads to the tendency to withhold or manipulate information gained through intuition for self-protection.

Seventh Chakra

This chakra lives above the crown of the head. It is associated with the causal plane, magic, and being able to co-create with the All Mighty One. The colors white and purple are located in this chakra. It is related to the pineal gland, upper brain, and right eye. When this chakra is open and the lower ones are more closed, witchcraft and mass manipulation of power can occur. When it is open and supported by the lower chakras, we can tap into the creative energy of the universe. We feel self-autonomy and control over our destiny. It is the gateway to transmutation. It is sometimes referred to as the "parking place" chakra. When it is open, we are aligned with Divine Will, and our wishes (such as finding a parking place) come true.

Visualizing the Chakras

When we view the chakras from the perspective of the etheric body (which looks like a web of gray-blue wire in the same shape as the body), they look like thin gray-blue funnels or flowers with petals around a central core. On the second or emotional level, they seem like clouds of different colors. On the third or mental level, they are perceived as thoughts or thought forms. On the fourth or astral level, they look like swirling pools of colored paint. Black spots in these pools indicate pain and negativity. The chakras in these forms are perceived with the Third Eye, which allows vision of higher vibrations.

Sounding Into the Chakras

Certain sounds from the human voice resonate with the frequency of different chakras and can influence the chakra itself. When we make a tone into a chakra, the vibration will feel and sound dissonant until the appropriate resonance is reached. This is indicated by a sense of harmony, which is felt by both sender and receiver. When the fifth chakra is open, this sounding ability comes easily and with pleasure. It has a profound effect on the vibrations of the chakras of the receiver and sender. Listening to a high-frequency chant or sounding a chant may have a strong influence on the energy body.

Opening the Chakras

The opening of the chakras takes place in non-linear fashion. There is not a direct progression of movement upward or downward. The movement depends on karma, destiny, and psychic resistance. The blueprint of the soul holds the pattern of this opening. Free will guides its process.

Message:

Hold close these patterns to your heart, for they are beacons in the night to guide you to the One. Know that within you is the exact degree of courage and determination to wash away all suffering and pain held tight within your being. Know that love beyond all else brings healing to your soul. Open unto love and all fear will dissipate. Lift unto the All Mighty One and sing the holy words of joy.

CHAKRA EXERCISE:

1. Sit quietly. Feel the support of the form on which you sit. Concentrate on your breathing. Follow your breath with your mind.

2. Concentrate on the place outside and the place inside where your breath changes from going out to coming in and coming in to going out. Focus your full attention on these points. Relax and breathe.

3. Concentrate now on the point within your chest where your breathing changes. Notice the space between the breaths. This is the space of the Self, of the All Mighty One, of love. Focus your attention here.

4. Now, focus your full attention on the base of your spine.

Notice the sensation here. Notice the emotion that is held here. Relax the muscles at the base of your spine. Let an image come to mind. Notice this image. Now send brilliant red light here.

5. Imagine the core of the Earth now, which is seen as brilliant red-orange light. Imagine a line of energy coming from the core of the Earth up through the soles of your feet, up your legs to the base of your spine, up your spine to your heart, and out your arms and hands. Feel this energy vibrate through you. Now imagine that this line of red-orange light becomes a flame that surrounds you.

6. Move now to the space in your pelvis between your pubic bone and your navel. Notice the sensation here. Notice what emotion is held here. Relax the muscles in your pelvis. Let an image come to mind. Notice this image. Now send brilliant orange light to this area.

7. Move up now to your solar plexus, just below your ribs. Notice the sensation here. Notice what emotion is held here. Relax the muscles in your solar plexus. Let an image come to mind. Notice this image. Now send brilliant yellow light to this area.

8. Move now to the center of your chest, to your heart area. Notice the sensation here. Notice what emotion resides here. Relax your muscles in your chest. Let an image come to mind. Notice this image. Now send brilliant green light to this area.

9. Move up now to the base of your throat. Notice the sensations here. Notice the emotion held here. Relax the muscles of your throat and neck. Let an image come to mind. Notice the image and send brilliant blue light into your throat.

10. Now move up to the center of your forehead, to your Third Eye. Notice the sensations here. Notice the emotion held here. Relax the muscles of your forehead. Let an image come to mind.

Notice this image. Now send brilliant indigo, blue-purple light to your forehead.

11. Now move up to just above the top of your head. Notice the sensations here. Notice the emotion held here. Relax the muscles on the top of your head. Let an image come to mind. Notice this image. Now send brilliant white and purple light to the top of your head.

12. Relax and breathe. Rest in the glory that is yours. Move down through the energy centers to find areas of tension. Breathe into these areas to release the tension.

13. Sit quietly in meditation for a few minutes. Gradually return to your body and the room. Write down the sensations, emotions, and images associated with each energy center. You may repeat this and discover different experiences. Each is for your learning and for your healing.

CHAPTER FOUR
HOW TO CHANNEL

Channeling is accessing the higher wisdom that is available to all who ask for it. It requires trust and surrender and a desire to open to this blessing.

Channeled communication can come through all the senses. However, most commonly we experience it through visual, auditory and kinesthetic modalities. The sensory vehicle for the channeling depends on which chakras are more open in the receiver and the vibration of the spirit guide.

Message:

Listen and you shall behold the Truth. Know that this Truth reflects yourself and all that you are. Know the channeling is your own Self come true. Know that all who wish to know the Truth will be able to vibrate with those sources that reveal the Truth. The obstruction is the fear of knowing that which you think you wish to know.

To open to the vibration deep within the soul is to reverberate with the sound of God, the sound of the soul, the sound of the Earth, the sound of your own wisdom, the sound of your own knowing.

We exist to speak to and through you, deep within your heart. You know we are here. It is the words that bring forth the vibrations already felt within your heart. The words themselves externalize that which is already there, glowing in the dark, so to speak. Vibrations from the tongue are the densest form of these words, for they exist in higher vibrational levels than the human ear can hear. Since all life lives as vibration held together by the blanket of love, all time passes through vibration. These words come through time to lie upon the fertile ground of your own being.

Hold your tongue and listen to our voices sing the praises of All That Was And Is Forever More. Turn inward to the wondrous song within your heart. Listen to your heart sing in all its glory. This is the goal of channeling as you call it. Listen to your own ring of flowers that surrounds you. Listen to the Holy One as the beauty vibrates forth to speak to you.

Allow the knowing to emerge within the golden splendor of your heart. Always love and blessings issue forth, flooding, bathing, nourishing the wounds encountered as the heart open to life upon the Earth and upon this plane.

The voices heard appear to be from different sources. For the mind, these exist to draw the curiosity to seek the source and merge with it. These are trumpet tones, beckoning hearts to open unto God. The entire purpose of our existence is to deepen love for God. Hallowed Be His Name. For God, the All, the Whole, the All Mighty One, exists as each and every sound vibration, each and every light vibration. All are intricately designed to speak His Name. Hallowed Be His Name. Hail forth upon this journey to the Hallowed Land of All, to the deepening depths of Love vibration, to the Glory that is All.

Know now, my children, that to know these vibrations known as channeling, is to know the sanctity of God's creation. Know also that these are to be used with utmost respect and Truth, forever seeking union with the Whole. Let love flow from your hearts and you will hear our voices. Love flows within and around to fill you. Love moves to the empty spaces resulting from the separation from the Whole, the destiny, in one sense, of all human beings. And so love comes to soothe the wounds, and the sounds and voices reverberate as love. Let time expand to open to the inner sound that exists already in its latent state.

Your guides are waiting to hold you and to envelop you with love. They take form in images that you can picture in your mind, but they are actually Light. They truly exist and they are there to help you discover your true nature.

Come with us to know your Self through lifetimes of experience, through travels to dimensions unknown on Earth today, through gradual, timeless torches lit to live for brief encounters in the flesh, to tread the albatross, to arise and be reborn.

In respect for the source of the sound and words, we ask that your love provide the foundation for the expression. To use the words for aggrandizement is to tread upon the soul. Hallowed Be His Name.

CHAPTER FIVE
SPIRITUAL GUIDES

Asking Questions

Because this is a relationship of frequency, the channeled response depends on the alchemy of those involved. The answer to a question will vibrate with the energy of the questioner and the channeler. Asking the same question more than once may result in different answers. The question may come from a different place of knowing or feeling within the questioner. The response will be appropriate for the vibration of the question. There are also varying perspectives on a situation. The response reflects the wavelength of the question and the inner knowing of the questioner. It is for the questioner to judge the relevance and truth of the answer. All is given to assist in soul growth, not to create dependency or to manipulate.

Interpretation

The channeling is usually pure; however, the interpretation of the information can be clouded by the channeler. In addition, the maturity and emotional well-being of the receiver will strongly influence the interpretation.

If the receiver has not personally released fear concerning the subject channeled, there could be a distorted translation. However, often there is a healing of the receiver as the information comes through to teach the receiver. If the channeling is pure, the receiver will feel an exalted sense of being, peace, and wholeness.

Dark Entities

There may be occasions when the channeled voice is not one of Truth. We can determine this by our emotional response to the tone of the voice. If the channeler feels uncomfortable and does not like the vibration of the voice, it is best to refuse to communicate with the spirit. It is very important to receive only that which is beneficial to you and your fellow human beings.

Some people may become enthralled with the darker entities who wish to speak. They can be seduced by the dark mysterious power of these voices. This is not recommended. Fear will attract dark entities, so it is best to release fear to avoid feeding and nourishing them. The attraction to meeting the darkness may originate in childhood abuse. In this case it would be wise to undergo transpersonal psychotherapy to release negative karmic patterns before beginning to channel.

The Goal of Channeling

Awareness of our inner wisdom and the presence of the All Mighty One are the goals of channeling. Channeling increases understanding of the Truth and enhances a sense of universality and interconnectedness. Spiritual guides are with us to protect and guide us and to be vehicles for the expansion of consciousness.

"Ask and you shall receive" is the underlying principle in channeling spiritual guidance. Fears may arise as this process begins, such as the fear of being overpowered, the fear of being consumed, the fear of intimacy, the fear of being abandoned, the fear of being wrong, and the fear of being judged.

It is best to begin channeling by meditating alone. You can ask questions in your mind and listen to the answers. Begin by asking one question and then writing the answer down immediately after coming out of meditation. This will help make the answers seem more real. The reception varies in terms of intensity, method and interpretation, depending on the life experience, which is recorded in the aura, and on individual karma.

Message:

We come in many forms and are always present for teaching. There are guides for specific purposes, guides for specific teachings, guides who have incarnated and guides who have never lived in physical form. There are guides for guides. It is through mutual decision prior to conception that one or several Inner Guides choose to be with each person through the lifetime. These Inner Guides may also be the Inner Guides for other people. The relationship is not linear or exclusive as in the human arena. It can be considered a relationship of the bonding of frequencies. Guides may change as the human being moves through the lifetime. Childhood energies change as adulthood flowers, and thus, the resonance alters and new energies may come to guide the older being. Ultimately, guides are present for teaching humans to expand in love and dedication to the Whole.

CHAPTER SIX
THE SOUL

The soul is the latticework of our being in motion through lifetimes. As essence, this form does not change. As substance, as genes clinging to the DNA, the form changes, adding, altering, and subtracting the items of substance. The essential energetic form remains the same. In this sense, there is no death. We just change form.

In the physical form, our chromosomal substances are influenced by tidal waters, climate, sound congestion, moon rays, planetary rays, and seasons. They are also influenced by the physical, emotional, mental, and spiritual bodies of our parents. Our siblings and others of close proximity influence our formation to a lesser extent.

At conception we draw magnetically a genetic coding of the five elements into our physical form, which directs the development of our individuality. This genetic coding is transmuted into our next form after undergoing processes between lives. Our coding designates the relationship of the elements in all aspects of the individual, with each incarnation striving for a balance between the elements on all levels. This creates a healing for the soul, which is continually focused and directed toward the goal of complete harmony. Soul growth within lifetimes influences this delicate balance. All healing of the wounds of our souls is ultimately dedicated to this harmony.

Deep within the chromosomal structure is a resonance with the One, a unity of vibration that is one and the same. This resonance provides the link between the chromosomal substance and the soul, acting as a bridge to join the two. As a stream flows among the rocks, this resonance flows through the chromosomal substance, interpenetrating the lattice of the soul. There is no boundary between these, but rather an interpenetration, as Light merges with Light. Each impacts the other in an intricate, highly organized fashion of creation—form within the formless.

The soul vibrates as a lattice of blue light, a less dense form of the etheric body. Upon this lattice the Self emerges. Colors gradually emerge to create the form. The soul is a vaporous essence. It is seen as a white or multi-colored cloud that flows within and through the denser form. The soul holds the memory of all experience in a high frequency vacuum. As consciousness balloons, awareness of the soul's experience expands. The essence of the soul remains the same even though it is altered by experience through time. It is possible to experience the soul as deathless being. There is no life and death, only being. This is what is meant by no death.

At physical death, when the soul leaves the body, the elements fold into each other and return to the lattice of the soul. The blue light that leaves the body at death contains the blue latticework of the soul, which has been imprinted with the experiences of this present lifetime.

The physical body, influenced by the emotional and mental bodies, assumes death is inevitable, and thus programs death to be. Through alignment with the soul, there is no death. Fear creates the expectation of death. This fear is programmed at conception, due to the beliefs of the parents and environmental conditions within the aura of the Earth. The essence of the soul is life and it is to this state that the entire being strives to return. Even in death there is the desire to return to the total state. There has been a clearing of certain energy fields, which now allow a greater motion toward this connection. Soul energy is emerging in consciousness to be known. So sit quietly within your own soul and be in the Light of your own Self. Know that which you desire beyond all else exists only to be known.

Your soul seeks to unite in frequency with your denser form, opening your heart to love, your own healing, and to the healing of planet Earth. This is so that you shall know the reality of your soul's calling. Be unto the One, the All Mighty One, and you shall belong to your own heart. Know that your soul awaits in full consciousness and patience for the return to unite with you.

    Open your heart to your truth and to your deepest desire to create. Upon the way you shall know the reality of your soul. The way holds forth for you to know peace with no fear. No fear exists in love, and in this meeting there is the deepest love. It is for your own soul that you hunger. Let love envelop you to heal the wounds and open unto the All Mighty One.

CHAPTER SEVEN
ARCHETYPAL IDENTIFICATION

How vast are the regions of human consciousness, and how deep are the chasms that divide. Hallowed be the Lord in name of all names that springs forth to enliven living in the human form.

Life energy transmutes to become forms within the human psyche. These forms hold contracted vibrational frequencies and take shape as gods and goddesses, gnomes and elves, ferocious beasts and whimsical fairies, vampiring bats and wistful waifs. They may come in the form of a hungry child, a wanton seductress, a ferocious beast, or a grieving aged man. They may horrify, delight, glorify, or diminish that which the human soul holds most dear.

These archetypes belong to the primordial source and become forms for the enlightenment of the human soul. These archetypes are a vital reality in the human soul life, for they propel the soul through lifetimes, torturing, taming, twisting, tempting. They fiercely, gently, lovingly, suffocatingly, bring into being the individual human being and civilizations of human beings.

The archetypes take life in myths, folklore, fantasy, dreams and stories that melt together to form the foundation for cultures and societies. Their presence is of a holy nature, enlivening, enlightening, firing the will to life, to the poetry of being.

Allow these archetypes to step into the Light. Light cleanses and purifies these energetic forms to create a house of fully enlivened beings, available for war or peace within the human psyche.

As these forms come to the Light, there is room for all. A magnificent energetic state of being comes into consciousness. Gradually, the beasts transform into human form, melting the primordial beings to become more evolved. The human forms become brighter and cleansed of darkness, leading to god-like forms of archetypal energies. These, in turn, transform to Light and purify the journey of the transmigrating soul.

Message:

Inherent in the human heart is a hunger to know. These archetypes stimulate and momentarily satiate this hunger. Each human heart has its unique conglomeration of these energies, which are carried through lifetimes in the vaporous substance of the soul. These are the genetic codes of the psyche, flourishing in full bloom in the enlightened being. This enlightened state comes as all archetypes come to the Light and flow in present time. The forms will vary according to civilizations and familial heritage, but the essence is the same. As each archetype strides forth in daylight, its full presence is known. Darkness fades and the individual energies are released from the twisted hold of the contaminated archetypal form.

Thus, a woman may hold within her the hero of a sage, an armed warrior, a sworded knight, a victorious king. And until she releases this male image, he may twist and turn against her in the agony of impotence. A man may hold in his heart a florescent, flowing maiden, who joyfully dances in the floral fields, fully knowing she is loved and beautiful. He may suppress her for fear of seeming superfluous and light-headed. She may, in turn, dishonor him by undermining his heroic feats. She may obscure details of war, distract him from the task, or prevent his attempts to master battle games.

My beloved beings, be in each and every archetype within your consciousness. Allow the flow. Appearance, reappearance, and disappearance. Bring to life these archetypal energies, to heal and lighten your soul.

Blessed be the variation in the human psyche, energizing the consciousness of Earth. Know that in lightening the archetypes, your essence is not diminished, but rather, glorified and enlivened. You do not lose your beingness. Instead, your way is clarified, your gifts revealed. Your heart melts to meet the All Mighty One and all is yours within your essential beingness. Freely flowing, the creative energies release the essence that is you. You, in turn, create upon the Earth that which is freed to come forth. It is a birth of the fullest form in the human realm. Know the glory that is yours.

CHAPTER EIGHT
ARCHETYPES AND SPIRITUAL GUIDES

Archetypes exist as healing agents to assist human evolution. A spiritual guide can be considered an archetype and be related to as such. Though a guide contains energies that go beyond the limitations of archetypal form, it is helpful to relate to a guide as an archetype. This assists in the differentiation of guide and Self, which stimulates a relationship of mentor or teacher. Often guides come in masculine form for women and can act as positive animus figures for women. For men this masculine energy can be a reinforcement of a positive masculine identity. In the feminine form, guides can offer a positive feminine identity for women and a positive anima figure for men.

It is important to note that true spiritual guides are always positive, loving, and wise. Magnetic attraction comes between the specific guide and specific human individual. This does not imply that there is only one guide for one person or that the guide is in one single form. Guides exist in an energy form at a much higher frequency than archetypes, although they can filter through in the form of an archetype while maintaining a separate identity.

ARCHETYPE EXERCISE:

1. This can be enacted alone or with a friend. Look at yourself in the mirror or have a friend look at you.

2. You or your friend can imagine another person transposed on top of you. Look closely at this person. Look at the clothes, the shape of the body, the expression on the face.

3. Now leave the mirror or separate briefly from your friend. Become this person.

4. Act out what you think is the nature of this person.

5. Reverse roles if you are doing this with a friend.

6. The person you have discovered is very likely one form of archetypal identification for you. You may wish to further you relationship with this person, developing a dialogue, so you can learn why it is there and what it has to teach you.

CHAPTER NINE
ALCHEMY

Alchemy is the art and science of transmutation where past, present and future are simultaneously known, experienced and loved. When the present is fully known, there is no fear. Faith in the All, evoked through awareness, carries the darkness away and leaves the golden light of dawn, each moment glowing with the light of love and God.

The alchemy of transformation upholds the principle that life itself is an alchemical process guided by Divine Will. It is timeless and ongoing and the purpose of human existence. The aim of alchemy, if an aim could be described, is the total Self, the unity of All. Healing in the true sense is the path to unity and wholeness. As water seek dryness, so alchemy moves to fill the spaces left empty by the human condition of separation from the whole. Alchemy resides in the soul of each human, seeping into spaces as they open. Alchemy feels, sees, opens, and heals as the transmutative energies vibrate through the fields of the entire Being.

Symbols will emerge as you ask. Seek and you shall find. Question not the meaning, but bless the beauty and wisdom of the images without judgment. They contain deep love and guidance for you. They come to bless your path and guide your faith with love. Love is the basis of all things. Love is real and true. Fear not the confusion of the mind. The higher order is not yet held within your heart. Trust in faith and love that these presentations have meaning. Trust in the divinity of All.

Alchemy manifests as change in the elemental structure of the body, thus the emphasis on metals and material substance in alchemical texts. The elements water, earth, fire, air, and ether are influenced by changes in the aura. As transformation takes place, subtle energies influence the physical composition and functioning of the body. Evolution within human consciousness and altering the vibratory rate of the body influence the elements.

Higher consciousness occurs in response to the opening of the chakras. The chakras are the link to the Universal Energy Field. They allow the utilization of Universal Energy, depending on the clarity of the flow of energy in each chakra. As consciousness expands through practices such as meditation, creative visualization, prayer, auric transformation, and chanting, alchemy occurs, transmuting limited consciousness to expanded consciousness.

Opening in the chakras is stimulated by karma—karma of past life experience and karmic consciousness acquired from experience in this lifetime and the karmic presence of the Earth during this lifetime. As consciousness expands, all levels of the aura are affected, thus influencing change in the physical, emotional, mental and spiritual bodies. As this change occurs, the structure of the cell changes to meet this vibratory level. There is a need for more minerals, for example, to sustain the change.

These alchemical changes, which can be seen with the Third Eye, indicate a deeper level of change. The person can come into direct contact with spiritual existence. Through transcendent experiences such as meditation, chanting holy words, dreams, visions of light, and movement of kundalini in the body, there can be direct experience of spiritual reality.

Transcendence introduces the presence of another reality, which allows the recognition of being more than ordinary, everyday consciousness. Alchemy provides the bridge between this everyday reality and the realm of spiritual Truth.

Times of intense alchemical transition are often accompanied by greater need for rest and water. Rest is needed to replenish the energy consumed by the change in physical structure. Water is needed to cool the friction caused by movement to a higher frequency. Emotional upheaval can create disturbances in personal relationships as emotional blocks are released. It is advisable to inform the partner, friend or workmate of these changes. Recognition of internal change can ease the possibility of fear or blame within the relationship. When there is resistance to these changes due to fear of change, there are often physical manifestations such as back pain, bloating and gas, headaches, kidney or gallbladder problems, colds and other symptoms of stress.

Message:

Alchemy is another word for love. Love provides the reason for the whole process in the first place—love of the Whole for itself in all its creation. So love is alchemy and alchemy is love. The alchemy of transformation is love transposed to the figure of the human form.

You are alchemy. You are the creator and the created. Speak not of alchemy as separate from what it is. There are varying degrees of alchemy. Alchemy is occurring at every moment of material and non-material existence. Alchemy is the thread upon the spinning wheel of life, and so is both life itself and that which is created in life form.

Alchemy is the most precious and vital elixir. It is the greatest of all elixirs, for it is the basis of all others. Alchemy merges your physical existence with your spiritual self. You are immersed in a profound alchemical process at this moment. You are blending and transmuting the four states of being without leaving the Earth, the body or the soul. As you open more to trust the true reality, visions and symbols will emerge from your own being, your own alchemy. You may also find mates for these among the ancient books. These are yours as you immerse your individual consciousness into the sea of totality and back again.

BLACK RAVEN EXERCISE:

1. Sit quietly and breathe normally with your eyes closed. Consider what you would like to create in your life. A certain income, a loving relationship, a fulfilling job, a house in the country, a loving circle of friends?

2. Allow what you want to create to come into focus. Notice the details of this form, which may be a symbol of what you would like to create. Notice the color, shape, texture, touch, sound and taste. Picture this in your mind's eye. Now let this image go, having firmly placed it in your consciousness.

3. Imagine a black raven. Look into the raven's eye. What do you see? This will lead you to the causal plane of reality, which is identified by pastel colors, geometric shapes, and pastoral scenes. Notice your surroundings. It is on this plane that alchemical change takes place.

4. Place your image on the causal plane. Acknowledge your worthiness to receive what this image symbolizes.

5. Now a fire comes and burns up the image on the causal plane. The fire's flames are the alchemical flames of transmutation. Allow the fire to completely burn the image so that nothing of it remains on the causal plane.

6. From the fire a silken mist arises, clouding your vision. Wait. Watch the mist. A rainbow will appear, signaling the bridge between the physical and spiritual worlds. Watch the rainbow.

7. Return along the rainbow to the physical plane. Let your awareness gradually return to your everyday reality and open your eyes. Notice forthcoming events. The time and specific manifestation of the image may not be quite as you wish; these are in the hands of the universe. Your desire will, however, be fulfilled. The "what" is up to you. The "how" is up to the universe. You have clarified to the universe what you want. Allow it now to do its work.

CHAPTER TEN
THE ALCHEMY OF SEXUALITY

Alchemy refers to the union of opposites as the source of the energy of transmutation. This is primarily viewed in terms of the union of the feminine and masculine principles or energies within and between human beings. In the merging of the feminine and masculine, combustion occurs, which creates change. This combustion, often symbolized by the sacred fire, transmutes one state of being to another of higher consciousness. Conscious awareness is altered forever. This change takes place in all the four bodies—physical, emotional, mental, spiritual—and may not be understood by the mind for what it is. The purpose of sexual union from the viewpoint of alchemy is to experience and transmute these energies to a higher state of consciousness.

On the physical level, sexual organs and glands are designed perfectly for the replication of the species. Each cell is consumed by the vibration of creative life force. In one sense, all cells of the body vibrate with the intention toward reproduction. However, specific cells are programmed more directly for reproduction. When there is energetic interference from other dimensions of the aura, the function of these cells is affected. This may appear as disease, lack of sexual response, inability to impregnate, diminished interest in sexual experience, and other physiological responses such as muscle contractions, heart palpitations, and momentary loss of hearing.

Within the emotional response in sexuality is the interwoven layer of emotional vibration, which pulses within the physical form. Emotions appear as colors in the aura, constantly spinning their expression of feeling. Since sexuality includes another being, even if only in the imaginary or archetypal form, all feelings related to another will be included in the sexual experience.

The other being most intimately and directly related to is the mother. The embodied child is held within the mother's womb, absorbing all levels of her experience of reality, including sexuality and sensuality. In essence, sexuality is sensuality, and thus the mother's sensuality in response to herself and others during gestation will directly influence the consciousness of the child.

The nature of pleasure held within the mother's womb and arms enters into the child's consciousness, deeply embedded as the experience of the other. From this initial vibratory impact, all other experience is added, blending and broadening individual consciousness. Each unique blending develops and becomes manifest as a composite of pleasure and pain, enhanced by the multitudinous variations of emotional experience. Thus, in entering upon a sexual experience, this full manifestation emerges.

Interwoven further in this perfectly designed being are thought forms which create expectation. Hovering below or above consciousness, these thought forms contain an understanding of the meaning of sexual (sensual) experience, both of the self and the other. Held deep within the heart is the knowledge of the One. It is the thought forms that cloud the comfort. As the sexual awareness rises, a brief glow of joy flashes, the memory of the One, the All, the Whole. This awareness may be covered quickly by overladen layers. Yet it is still known, even if temporarily forgotten. The hunger for this joy, this ultimate pleasure, continues, thrusting stronger and stronger until the explosion of the orgasm or until the fear overcomes the thrust and the hunger contracts in fear of anticipated pain.

Within the individual experience of pain and pleasure is the consciousness of the collective—the force of thoughts pervading multitudes of time. These forms enlarge individual thought configurations. They impact with full force the consciousness of time and space within the present cultural milieu. This impaction seeps deeply into consciousness, creating a further distortion of true reality. Consciousness of power, safety, and security within the group erupts to impose definitions and expectations of the sexes, darkening the clouds that veil the One. Rather than with open arms in joyous pleasure, the sexual meeting is approached in clouds of darkness, filled with fear.

The Goal Is Union

The goal of alchemy is the union of the masculine and feminine energies. This is represented by the symbols of men and women in sexual union. On a physical level this is symbolized by the Philosopher's Stone, the elixir or tincture which could turn lead or base metal into gold. The transmutation occurred as the metal turned to gold, but the power was in the stone. It was considered to be the transmuting agent. The stone, which is said to be of the nature of gold, is able to accelerate the process of perfection in the material world. The Philosopher's Stone was thought to be more perfect than gold.

This union is the true nature of transformation and the purpose of living. Sexual relationships offer the opportunity to open to Light through the experience of the union of opposites in physical form.

The theme of alchemy is that true transmutation comes through death and rebirth. Alchemy involves the process of moving through the blackening stage, when there is the meeting of two opposites. It is associated with death and the descent into the dark chaos of the unknown. The whitening stage occurs when the spirit of the matter ascends as the male and female energies merge and feel the initial joy of union. This is followed by the yellowing stage where identities are confused and there is resistance to change. The final stage, the reddening, occurs as the two unite in a new manner, grow to maturity, and create together.

Relationships of all kinds incorporate these alchemical stages. These stages can occur instantaneously or over a period of time, depending on what level of interaction is being expressed.

In a sexual relationship between a man and a woman, sexuality from the alchemical standpoint can become the means to attain God. The process is one of transforming sexual energy to Divine Consciousness. From the beginning of consciousness there has been a dynamic tension created from all parts seeking to unite with the Whole. This creative tension is the constant magnetic drawing together and the opposing force of pulling apart. Thus, the two polarities move back and forth, seeking to join. Yet they fear loss of identity, and so they draw apart.

From an energetic scale, the first and seventh chakras are in constant tension, the first drawing to the earth and the seventh drawing to the spirit. These two poles manifest together in the spine as a pole (the staff in the caduceus), and the masculine and feminine energies encircle this pole, climbing to meet at the top (the snakes in the caduceus). In the Hindu paradigm, the central cord is called the sushumna. The masculine and feminine cords encircling it are called the Ida and the Pingala.

In the human body these poles manifest as genital organs. They are perfectly designed for sexual union and are waiting for conjunction with the opposite. The density of physical matter prevents true unity on the physical level, although the hermaphrodite is a physical expression of this union. In most humans there is the seeking of this unity outside the individual in the form of sexual union. In intercourse there is union for a short time. In the parting there is a renewal of the creative tension. However, within the physical union, conception can take place in a perfect state of creation.

Sexual transmutation can be accelerated by opening the chakras to receiving nourishment and stimulation from the Universal Energy Field, leading to deeper and more fulfilling sexuality. The sacred fire spoken of by alchemists is the fire of transmutation, the combustion that occurs as the opening to the Universal Energy Field takes place.

There is a need for moral maturity to parallel sexual development in order to avoid the misuse of this potentiality. By opening all the chakras, particularly the heart, a balance in energy prevents this misuse. This is the primary reason that alchemy has been kept secret for so long. In the opening of all the chakras, there is diminishment of sexual repression and a melting of the projection of fears that cause pain and misuse in sexual relationships. The purpose of alchemy and sexuality is to reach God. Any distortion of this reality is a distortion of Truth.

As spiritual development progresses, the cellular structure transforms so that it can utilize the energies of higher frequencies. The nervous system must be able to hold the vibration. This occurs most harmoniously when the physical, emotional, mental, and spiritual bodies are in balance. Fear weakens cellular metabolism because the energy is expended in the fight/ flight response to meet the anticipated enemy and/or arising (death), rather than to reach the higher level of the trance state. To remain at higher vibratory levels without the release of fear is a detriment to health.

Sexual energy is divine creative power, vibrating and pulsating within each human being. The Hindus portrayed this energy in their erotic coupling of Shiva and Shakti and the symbols of yoni and lingum, which appears to be a penis growing from a vulva. This union is referred to as the Mystical Marriage. These symbols are worshipped throughout India as Divine energy.

Alchemical sexual development must be directly experienced. Reading, imagining, and conceptualizing these stages is not enough, although this is helpful in causing transmutative forces to create change. If sexual energy is not experienced in actuality and in relationship, there is an imbalance that can create loneliness, fear, repression, and ill-health. These evolve from the pressure to integrate the imagined experience into reality.

Message to the Woman:

Allow the fire in your loins to grow and burn with the intensity of the wind across the desert. Breathe into this burning heat. Growl from the molten lava between your legs. The glory of God resides there. Beneath the pain in your heart is the hunger in your groin—the thunderous hunger to meet the

God. Revel in the glory of your passion. As woman, you know the passionate cry, "Oh Lord, make us One!"

Beneath the branches lies the treasured jewel, adorned and anointed as a queen bee amidst her honeycomb. This beautiful jewel, this glorious cone of pleasure, is the key to a most perfect gift. Cry, my beloved, for the flames of love sear away the hardened layer of fruit to uncover the nectar of truth. For it is truth that is this gift, the truth that you and I are One. The fire is the truth, for it is here that the burning love bursts forth in all its holy heat.

Message to the Man:

You, too, hold this fiery love, the chariot of fire. Yours is of a different nature and is as glorious. It is simply held in a different vessel. Your fire moves out while the woman's remains contained within the vessel. Your shining sword of passion burns with brilliant heat, penetrating darkened shadows to reveal the light.

Deep beneath the silver lake of the calm facade of your being lies passion—passion long contained within the deep, dark dregs of fear. Remove the stone that holds back the flood. Let the water that so cools the fiery heat flow forth, released, unheeded. Tears and fears and desires long kept hidden gush from the mouth of God. Thus freed, flames leap to meet the laden air with pure pleasure of the meeting. The passion is released from the bondage held within the energies of the soul. The fire holds the treasure of the soul. You are one. The fire in your groin glows with the intensity of your love. Man and woman meet in perfect harmony and heat—I AM, I AM, I AM.

The heart holds the water dammed by the stone. As the water is released, the heart ignites the fire in the loins. As the heart feels, the water flows and the heartache opens to meet the face of God. As a tree unshadowed grows to meet the sun, so the flames now released by the watery flow leap to lap the sky. Man and woman, so ignited, burn in the glory that is God's.

As spiritual hunger increases and the heart opens, the way will become clear. All that is necessary is the desire to know the All Mighty One. This desire will lead you to these experiences. Open to the energies of others who wish the same and you will come to know and trust your divine judgment. You will come to welcome those partners who will enable this growth. Ask in your heart, *"Is this action, this sexual embrace, that which I desire in truth?"* You will be led by your own heart to the partner you so desire.

SEXUAL ALCHEMY EXERCISE:

1. Sit quietly, naked, cross-legged, facing your sexual partner. Both close your eyes and open your chakras. Meditate for ten minutes. Take several deep breaths.

2. Open your eyes and look into the other's eyes with full awareness of the meditative state. Imagine the other as a god or goddess. Describe what you see to each other.

3. Say to the other, "I worship God within you." Hold hands, continuing to be aware of the opening of all chakras.

4. Begin to focus on grounding red-orange light to the center of the Earth. As soon as you each feel this grounding, open further the first chakra and extend this energy to the other, joining your first chakras.

5. Continue upwards, opening and joining the second, third, fourth, fifth, and sixth chakras. Sit in this union.

6. Now begin slowly to make love as you wish. After you both feel you have reached the end of the sexual intensity and there is the physical separation of the sexual organs, continue to feel the opening and joining of your chakras as you lie quietly together. Experience the nature of your union at this time.

7. When you decide to get up or go to sleep, place your hands together. Say to yourselves that your partner has the power to heal himself or herself until you feel the release and believe it to be true. You may repeat this exercise as often as you desire. You may wish to note the differences in your experience as you evolve in this union.

CHAPTER ELEVEN
PROJECTION

Projection is the extension of the perception of one's unclaimed thoughts and emotions onto another. To experience the power of perception, look inward to the taming of the sea of lights within. Look to the dragons, demons, angels, devils. Look to the child, the old wizard, the perfumed dancer. Look to the sky diver, the coal miner, the dancing bear. Look to the myriad of forms and swellings within your underworld.

Projection seeks its counterpart outside to give it life. Without this external form no blood can flow, no air is breathed within the projection. As an octopus reaches out to suck in nourishment, projection prowls upon worldly soil in search of nourishment. As the cloth moves across the mirror to clarify the forms reflected, images will clear and reflections will lighten.

There will always be projection, for it is the nature of communication. The nature of projection alters, however, with the cleansing of the glass. The ingredients of projected light are multifold: karmic, developmental, archetypal, ageless. The rainbow of light flows within and without, carrying with it the remembrance of these energetic images, forever moving to enliven All That Is.

Projection is a universal vehicle for developing individuality within the context of the Whole. As understanding and interpretation of the external world evolve, images are reflected outward. What is experienced internally is thought to be true of the external world. If there is a sense of badness inside, this will be projected outward, and the world will be a bad and fearful place. If love and goodness predominate perception, then the world will appear as Light and love.

Projection is the mechanism through which you see the

external world. It is the mirror of God's work and is often misinterpreted as harmful. Projection is included in spiritual guidance and is a vital aspect of this process, for it grounds the illusory in human form. In order to relate to God or spiritual guides, it is the human tendency to create a form on which to focus. It is very difficult to imagine God or guides as sheer energy, so the imagination creates forms for them. These forms are projections of what they are thought to be. They can be seen as projections of the Self.

As the Self comes to be experienced as holy, receptivity to spiritual projections intensifies. The love of the guru or transformational guide or lover or God is taken inside and becomes a positive reflection of the Self. This then is projected outward onto other people and spiritual forms. The love for others and for spiritual forms increases the love of Self.

Claiming negative projections assists greatly in empowerment and healing. The energy expended in containing the projections can be directed toward self-understanding and creativity. Projections result from the fear of feelings within, such as the desire to leave a relationship, a desire to be taken care of, a desire for someone of the same sex, or a desire for material pleasures. All these can be projected outward through believing that others are functioning in the same manner. Once the truth is known that these are desires within, a decision can be made as to how, or if, to manifest them in action. Once these are revealed and claimed and brought into the Light, the power of this energy is diminished. There is more choice in action.

Projection can be a valuable mirror for uncovering the Shadow and releasing the beauty within. That which you wish to cleanse is but a darkened form of your true nature.

Message:

When Light meets Light, there is only Light. Thus the projection merges Light to Light to Light, ongoing and complete. As you listen to the

God within, hail the Hallowed Name. From within all glory shines. It is to you that the torch is passed to seek the Light within. Find the dark place within where the Shadow lurks. Outline its form, its name, its speech, its movement. Move it to the door of Light. Open the door. Lead it forward when no earth appears upon which to stand. Step into the amorphous, spacious sea of Light, hand in hand until the Shadow melts away. There is no abandonment, but rather, letting go of fear. This is the nature of human evolution, which allows the transformation of fear to Light. It is an internal form bursting forth in the beauty and glory of creative thought and vision, the source of human evolution.

I am of you, of your wisdom from ages past, of your knowledge gained within this lifetime on Earth. However, I am also the voice of holiness, the angelic form of the Word of God. Seek not so much to understand, but to know the reality of the words, to absorb deep within your core the truth of what is known. As you listen, let the sound blend and melt away the fear. Let the eyes of the Truth beam Light to all the darkened holes, deeper and deeper until the Light and Dark are one, and no fear exists. Know this is possible and real.

PROJECTION EXERCISE:

1. Sit quietly. Focus your attention on a relationship in which you are not comfortable.

2. Look at your two fists. Select one to be you and the other to be the other person.

3. Imagine that you are the fist that you have chosen to be you. Tell the other fist (person) how you feel about him or her. Let the other person tell you how he or she feels about you.

4. Tell the other person how you felt about what he or she said to you.

5. Tell the other person what you want from him or her. Let the person tell you what he or she wants from you.

6. Each of you respond by saying how you feel about this request.

7. Each hand represents two important parts of yourself. Each wants recognition and attention.

These feelings are likely to be real in the other person as well as in yourself. Consult gently with the other person about the reality of what you have discovered and about how they feel. Ask to understand and do not judge. There is no right or wrong in projection. You will find that the more you explore the projection, the more it comes home to you. As you become able to claim your projections, you will feel the power of choice and acceptance of your own reality, both Shadow and Light.

CHAPTER TWELVE
TRANSFERENCE

Transference can occur in worldly relationships, between man and woman, parent and child, employee and employer, friend and friend. The energies of transference are easily ignited in an authoritarian or romantic situation, fueled by the tendency to enact early childhood dramas. Expectations and fears of reliving past relationships distort the truth and potentiality of meeting in love. Thus, the healing of transference is most beneficial to the healing of the whole being.

Transference is usually activated in a transformational situation, whether the contract to transform is formal or informal. The implicit or explicit agreement is that one person will help another. This revives earlier helping relationships, particularly that of the parent and child.

The drama of earlier relationships is held in the unconscious in the emotional body. Until these dramas are re-enacted in such a way as to heal the past pain, they will remain dominant forces in the emotional life. When the early relationships are positive and loving, the internal dramas are played out easily and with a sense of fulfillment. If the essence of parental love is weighted toward trust and loyalty to the child's positive growth, the child will learn to expect the same in other intimate relationships. The internal drama will continue to be loving and growth-producing. For example, the child will draw a lover who also had loving parents and will develop a positive loving relationship with that person with little conflict. However, due to the nature of karmic interplay, these dramas often hold negative energy configurations which are present in order to heal karmic pain.

If transference is positive and enhances love, no investigation is necessary. However, when there is obstruction to receiving or giving love due to transference, interpretation is helpful. It is to lift the veil of transference to discover the reality of the image behind the veil. In the lifting of the veil, the image discovered may not be one of a kind loving person, but will be the reality of the relationship. When this reality is faced and absorbed, the attraction to transferential objects who play out the negative transferential relationship dissipates.

Message:

Transference is the heart within the heart, the womb within the womb, the soul within the soul. It is through transference that doors open unto the Self that is hidden beneath layers of personal identification and projections of distorted visions of the Self onto the outer world. Transference feeds the Hungry Ghosts—ghostly forms of projections inclined to feed upon the energy of the soul. Through bringing Light to transference, the Hungry Ghosts are ingested and consumed. One can let go of the transferential bonds that bind the soul's tasks on Earth.

Transference hides the holiness of each being and each life. Transference chains the soul to karma and can only be recovered by feeding the Hungry Ghosts—layers of personifications carried by the soul in a lifetime. They demand attention that could otherwise be directed toward soul growth. They carry illusions about the Self in relation to others, expectations of being the sacrificial victim, the abhorrent whore, the dejected warrior, the wanton wailer. They appear within the auric form as images of hungry beings demanding attention. The form may be demonic, ghostly, grotesque, bewitching or beguiling. Each form represents the hunger to be fed.

To feed them is to first acknowledge their reality, and second, to surrender deeply to the intense hunger, thus freeing bonded energy for growth. The Hungry Ghosts are universal in human consciousness and yet are unique to each human being. They are known as archetypes and internalized objects. They seem as demons persecuting, causing pain and fear, limiting the growth toward the All Mighty One, blinding one from Truth. They appear as ravenous, demanding, engulfing, absorbing. They are the transferential realities as exorcised in the healing of the mind, body, and spirit.

These originate in the conscious mind within the parental childhood bond, which passes through generations the fears and limitations of the Self. We choose parents to provide the perfected transferential ingredients for growth toward the glory that is God. As we meet the Hungry Ghosts and confront within the deep conviction of one who longs to know the truth, the Hungry Ghosts will melt away. What remains is Light.

To heal the Hungry Ghosts, first acknowledge their existence in every aspect of life, how they orient the Self to love and fear of love. They exist within the auric field as blocks of energy in the second layer. They hold the power of the love of God, but they are compacted and contained in the fear of separation from the Whole—the fear of loss of God. In transforming the emotions, cries of longing, long repressed, spring forth. The Hungry Ghosts are fed and dissolve into the Light.

As longings are released, love pours forth to soothe the wounds and the fear, transforming them to love. Love honors and respects the Hungry Ghosts, which inhabit your being until they are dismissed. As love for self and other emerges from the clouds of molten energetic form within the aura, the Hungry Ghosts are met. As compassion for and acceptance of fear (which enfires them to battle) expands within the human heart, the love for Self and other expands in direct proportion to this growth. To fear the fear is to give power to the Hungry Ghosts. To love them, and to acknowledge their fear as a projection of the fear held so deep within the human heart, is to heal them and so to heal the human heart.

Open to the healing of the Hungry Ghosts through love. Love the fear held within your heart. Cry tears of deep compassion for hungers that go denied. Love from your heart will wash away the wounds. You can feel the purity of love in infancy before the hunger grew so great that it threatened life.

The aloneness you may feel as layers of transference are shed from your shell and shatter the aloneness of the soul upon its journey. Transferential issues that are now manifest as love, truth, beingness and relatedness begin anew. Parental ties are loosened and transformed, and there comes a time of wandering as lonely as being alone in the desert or on the sea. As the pain of long ago is released unto love, there may be the sense of emptiness and loss. Surrender to this loneliness, this emptiness, and you will find a deeper love, the love of your own Self.

TRANSFERENCE EXERCISE:

1. Sit quietly in meditation and allow yourself to imagine a face. Select the first face that appears to you.

2. Notice how you feel about this person. Communicate silently to this person about how you feel about him or her. Let this person respond to you as to how he or she feels about you.

3. Now, allow yourself to sink backwards in time to when you were an infant. Imagine your mother as she was at this time. Look into her eyes. What do you see there? Notice how you feel as you look into her eyes.

4. Imagine your father at that time. Look into his eyes. Notice how you feel as you look into his eyes.

5. Imagine another important person in your life then. Look into their eyes. Notice how you feel as you look into their eyes.

6. Now return to the face of the person you first saw. Notice if there are any similarities to the eyes of the person you have just seen. Tell this person silently if they remind you of any of these three people and why.

7. Ask yourself how these eyes influence your interaction with this person. Ask the person how you might feel closer and listen to the answer.

8. Say good-bye to this person for now and return to your normal consciousness.

CHAPTER THIRTEEN
TRANSCENDENCE

Transcendence is the process of expanding reality beyond the normal ego identification with the individual Self. All spiritual practices such as prayer, chanting, singing, contemplation, meditation, dancing, and ceremonial rituals facilitate transcendent experience. Transformation occurs as transcendence interweaves with the working through of transferential relationships. In the struggle to find freedom from the imprisonment of transference, transcendent experiences lift and expand to ease and guide the way.

Message:

Surrender unto yourself. Experience those worlds within that can emerge as daily reality is released. Know that the transcendent experience is but another dimension of your own reality, your own Self.

There are those who seek to escape the demands of daily life by moving into higher dimensions, and there are those who cling to daily life for fear of being lost in the higher dimensions. Neither recognizes the call of God within—deeply, completely saturating the entire being. Transcendence is the face of God presented to you as a summons to bring you ever forth to know the Lord within. Know that to be seduced by God is the most holy surrender, a surrender unto your own Self.

Sink deeply now within. Surrender ever more to the being that is you. Allow the messages to enter as they do, for you alone, and bow in love and wonderment in the knowledge of the Glory that is All.

Lift unto this Glory through the source of sound, of light, of speech, of contemplation, of visionary apparition. By chanting sounds that reverberate within to align the chakra energies, the experience of transcendence can reach wholeness. Light bursting forth in consciousness, in a dream or meditative moment, can bring awareness of the All Mighty One within. The tone and meaning of the words that inspire Truth, that bring realization of the Whole, create a transcendent recognition of the reality of higher consciousness.

These words flow through centuries of time, the Truth transcending place and person, washing all that listen. In contemplation, words cease. And in the stillness, God is known. The silence from within, in presence of the Whole, knows constantly the Truth. This contemplation underlies the foundation of the Truth, and all who know this statutory state are beholden unto God. In visionary form, reality detours from the trodden track, and wondrous sights emerge to lead the way. These visions may be of Holy Ones, of teachers long ago, of parents moved to the arising plane of being, of children so endeared. These visions are reminders of consciousness expanded and the Greater Truth within.

Transcendence moves within experience alone. The transcendent truth emerges within a sensational form in order to be comprehended by the one who experiences the form directly. These words may bring you inspiration, but it is through your senses that you will come to know this Truth. The senses of the worldly being are thus expanded to encompass higher frequencies of sensation. Thus, transcendent being occurs through the normal sensory experience, expanding to reach a higher frequency of being. Your eyes will see beyond the normal sight to inner vision and to light. Your ears hear tones unknown to the ear of normal consciousness. These tones can stimulate the higher frequencies themselves. Thoughts of the All Mighty One bring joy and love and elevation to an altered state of being. It is through recognition of the reality of these transcendent experiences that the hunger to know the reality of the All Mighty One is drawn.

Transcendence is best developed within the context of a contained environment, such as the hallowed halls of a spiritual temple or ashram, the security of direct contact with a master, the safe office of a psychotherapist, the warmth of a healing circle, or the arms of an experienced friend or lover.

An important step toward transcendence is the willingness

to explore other modalities of sensory awareness. This may come as a drug experience (not recommended, but sometimes an introduction), a visit to an ashram or temple of holy light, hypnotic trance, biofeedback, meditation, reading holy scriptures or being in the presence of a spiritual master. In Western culture there is an aversion to acknowledging the paradigm of wholeness in preference for that of mechanistic science. Otherworldly pursuits may be considered demonic, irrelevant, dangerously seductive, unrealistic. Courage is required to break through this common belief system in search of a higher truth.

A church, ashram, or other holy place is often associated with mother. Many fear entering a place of spirituality in the same way they would fear becoming close to their mother. If your mother was suffocating, humiliating, engulfing, or demanding, such will be the transference onto the place of worship. These feelings are compounded by the mother's attitude toward religion, such as over-investment in the church's protective ability, or distrust of authoritarian institutions.

Transcendence is a release of everyday reality to expand to higher consciousness. There is a letting go of mental constructs and a melting of ego boundaries in order to open to transcended reality. When this is done with conscious choice, fear is lessened because there is the sense of being in control. Transcendence brings truth directly, magnetizing the path toward transmutation. Transformation takes place as the transference expands to include the All Mighty One.

CHAPTER FOURTEEN
TRANSFORMATION

Transformation moves a human being toward the Whole in a step-by-step process of unfoldment. One looks back into the dark spaces that have been alienated from the Self as fragments that have been lost are recovered.

The instrument that guides transformation is transference, the internal vision governing perception of Self and the world. It is as if two-sided glasses are worn, both looking inside and outside, but the glasses may act as a veil to block the Truth. Removing the glasses, such as during times of transcendence, reveals the Truth beyond the veil that you and I are One. Transference separates one from another in that the Self is seen as reflected in the parent's eye, and so, each human's reflection is an individual perception of that which is seen in the parent's eyes.

Message:

Listen deeply to your heart, and you will find a voice of love, the love so longed for as a child. Know that the only true desire of a child is to see this love shine forth from the parent's eyes. Instead, so often, anger flashes, or sadness bores, or fear shivers, and the child knows only this. And yet there is the Inner Light, shimmering, though sometimes faint, that reminds the child of love. Thus, as the child grows to maturity, these co-exist, vibrating together, yet at a different rate. The parent's eyes are taken inside and become the Self that is known. The child seeks to know and hold the inner Light in the face of the cold dark night, clinging to this Truth—felt but as yet unknown. The degree of darkness varies, but all humans know the darkness.

Transformation is turning up the flame of love to melt away the darkness. As parts of the Self so long hidden come forth into the Light, trust of the Greater Self emerges to fill the void. Restraints are lifted, and creativity rears up as a hungry dragon, so long starved for milk and honey. Rest now in trust that all is well and that the Truth shall set you free. Light will seek the darkest corners. Sink deeply, deeply into your own heart and know the Light is there. There can be no greater gift than love.

Transformation occurs as the integration of transcendence. Transference burns away the fear. Transference moves in a step-by-step progression, though not linear in design, to remove the veil of projection. Clarity is gained as the experience of parental influence is separated from the inner experience of Love and Truth. The transferential projection may not only be from child-parent experience but can include relationships to siblings, teachers, grandparents, aunts and uncles, cousins, and friends.

Transferential blocks can be seen in the aura as brown or dark-colored blotches of light. As the blocks are released, auric consciousness lightens and a new understanding of reality is gained. There is often tremendous resistance to the release of transference during the transformation process, for until another reality is known as true, the transferential relationships are the only sense of Self and other that the human knows. Therefore, it is essential to intersperse the working through of transferential issues with experiences of transcendence.

Chanting sounds of the All Mighty One carries the reality of the All Mighty One within. Immersing in the vibration of the sound, uniting with the sound of the voice, blocks the emotional attachment to the transferential relationship. The vibration of the ancient chants or chants from a state of guidance directly influence the energy in the chakras, changing the vibratory level of consciousness. The vibration stimulates realignment of the energy flow, breaking through the blocks in the auric field.

Sound can also be sent directly into another person's chakra through sounding with the voice in combination with energy transference. Sounding, if administered in a transcendent state, will correspond perfectly to the state of both beings, and is recommended for many transformational situations. The sounding will provide an opening in the energy field. The one who is sounding will know its effectiveness by the profound pleasure and empowerment felt within.

The goal of transformation is to realize the Light within. This realization gradually penetrates the aura, affecting all levels, lightening the darkness to reveal the Truth. Each person walks along their own individual path with Free Will, moving toward or away from opportunities to expand and deepen the experience of uniting the Spirit and the Earth within. Transformation opens the way to transmutation, the embodiment of this unity and the release of the soul's potential. The purpose of transformation is to open the door to the enactment of the soul's potential on Earth, the contribution toward evolution of the Whole. The tide turns up to the joy of union with the Divine and back down to Earth to the manifestation of the soul's potential.

TRANSFORMATIONAL EXERCISE:

1. Consider a relationship you would like to transform. Imagine the other person in the relationship sitting before you. Notice the details of the face and body.

2. Focus on your chakras, becoming aware of the color, tone, strength, tension, and speed of vibration.

3. Now look at the other person's chakras in the same way.

4. Come back to your own and notice where you feel tension and blockage.

5. Send healing energy into these places and release the blocks. Observe the effect of this release on the other person's chakras.

6. Surround the other person in gold light. Now surround yourself in gold light.

7. Open your eyes and write down whatever comes to you about how you could heal this relationship.

This can be done with a spouse, lover, parent, child, friend, teacher, or work associate. What is important to note is that by changing the energy in your chakras you can influence the relationship in a positive manner. Observe how the relationship changes.

CHAPTER FIFTEEN
TRANSMUTATION

The aura exists as a series of concentric circles of light vibrating within and in sequence with one another. The inner circle is the density of the physical form and the outer circle is the tenth layer of the aura and beyond. Transmutation is the fluid movement between these layers with the ability to align one's will with Divine Will, thus creating movement between the layers at will.

Basic to transmutation is the assumption of potency within as the Light of the Self. From the point of view of the human, potency may be distorted to form aggression or control over a fellow soul. The inner potency reflects the Inner Light, which shines with purity and love. It is the interpretation of this Light, as it shines upon the minds of those who grow in fear, that creates the concepts of danger of inner potency. Thus, to shine forth in love and honesty is frightening to those who have learned to interpret this force as threatening.

The threat sears the heart in the other who receives this Light, for there is a deep reminder of what is missing in the knowingness of the other. In the observer a longing stirs, the longing to know God. Simultaneously, the heart tightens in fear: the pain of desire unmet constricts and twists the heart. Best, it believes, to destroy those who live in the Light than to be reminded of this pain.

The awareness of this constricted attitude can affect the one transmuting on a physical level. This reaction to the Light can ignite fear within the one who shines in the Light, fear that this exposure will bring annihilation. At the same time, there is the desire to release the potency in all its glorious colors. There is a vulnerability that precedes the magnification of Light as it emerges from behind the clouds of negativity. This vulnerability may manifest as bloating in the stomach, rapid bowels, shivering, digestive problems, muscle tension and other symptoms of stress.

Message:

Seek God, and you shall find the World alight with the glowing presence of the All Mighty One. Transmutation is Light meeting Light; the frequencies meet at the same intensity. The internal and external vibratory rate are the same. Thus, let the fire of the Earth burn and you shall know the meeting of Air and Earth and face in all your brilliance the eyes of the Lord of All, the One.

The initiation of transmutation begins by clearing all seven layers of the aura so that one sits in clear red, orange, yellow, green, blue, indigo, and white-purple light. As the vibrations increase, clouds of the colors of the higher chakras increase around the head— blue, indigo, purple. They gradually become all white, flowing through and around the physical body. From the third chakra come bubbles of clear thought and clear etheric energy into the universe. The thought is not in verbal form but s pure mental energy.

This method of healing influences the etheric energy of the planet while at same time cleansing the third chakra of the healer. Like clear water rinsing a wound, no darkness can collect as the movement occurs. Eyes beam red, blue and yellow light when stimulated to do so, appearing as flashlight beams of brilliant light. The energy is Light—clear, non-emotional, and yet related. The distinction between those transmuting is clear and respectful and enjoyable. The differences between them are heightened but there are no neurotic ties to constrain or manipulate. Thought forms emerging are clear and pure.

The bond of love is there for all. At the same time it can exist as a loving bond between individuals as a karmic fit, a karmic partnership. This bond is similar to two pieces of a puzzle fitting together. There is a uniting and an exclusiveness, a magnetic unit, as the two are drawn together as natural partners, flowing with each other.

No one is truly alone, but some float, never firmly uniting or fitting with another, touching one and then another. Some bond in sequence, some have different bonds with different partners.

Transmutation can occur within a family. Children can be raised with full appreciation of the mutual choice of incarnation of the child and the family. With this awareness parents realize that they and the children are together to learn and to grow, assisting each other through the opportunities presented in their life together. They understand the uniqueness and the beauty of each child. They know they have found their perfect partner for helping both parents and children to transmute their karma. The children will grow in the light of love and guidance toward their uniqueness. In this process all souls involved find expression easily and each personality flourishes in its full expression.

The initial step toward this transmutative process must come from the parents in clearing their energetic fields. Once this state is reached co-operation occurs spontaneously. They accept the uniqueness and valuable contribution of each person, which is vital to the functioning of the whole family. The governing structure of the family can arise cooperatively with decisions made from a holistic perspective where no one is losing, all gaining. There is enough for all.

The meeting of two souls in love, laden with karmic darkness, is not a partnership of convenience or of superficial love. It is a bonding for the purpose of transmutative growth. This is a time for each of the two to end the self-blame and dig the hole deeper into the soil to find the hardened obstacles that prevent the love from flourishing. There is often a wish on the part of both that the other change to meet the need so defined, rather than the willingness to look within.

This relationship is like the mixing of the ingredients for a cake. The unmixed and uncooked ingredients do not create a cake. The cooking is needed. Attempts to wait or asking the other to change prevent the immersion in the pools of darkness that you must enter. It is time to share fears and open your heart to the humanness of the other in trust and devotion to your love.

During times of intense healing and change there are physical and emotional comforts that can be nourishing to the wounds as they open to be healed. These include drinking many glasses of water a day, deep sleep, gentle and loving relationships, remaining in the same dwelling place, nourishing food, sunbathing, salt water, massage, healing and helping others.

In times of adversity place a blue shield of light upon your front and back, wearing it as invisible armor against attack or manipulation. Activities in the fresh air and sun that involve moderate exercise are recommended. Also recommended is sexual experience with a slow rhythm, building very gradually to a crescendo. Sharing love with family and friends in intimate gatherings, surrounded by love and external security, is most beneficial.

CHAPTER SIXTEEN
STEPS TO TRANSMUTATION

Steps to transmutation are the states of being that occur as the human being moves toward living in the Light. They come not in direct sequence but rather in perfect order for each one. The earlier steps are traversed most often before the later ones, although the later ones are often touched and tasted as the earlier ones are known in depth.

ONE: Searing the heart to bring to the surface distortions of the Truth.

This may occur through psychotherapy, self-explorations, or another form of spiritual healing, bodywork, recovering from trauma, near death experience, spiritual practice and study and/or self-revelation.

Many in the Western world may not venture as far as this. Identification with the ego is strong and there is little exploration of the nature of Self. Moving into darkness, or the Shadow, to discover the reality of unconscious beliefs and memories, brings pain to the surface. However, entering into blocks in the physical, emotional, mental, and spiritual layers of the aura will help release the fear that obstructs awareness of the true Self. This is the first step into the unknown, into the unconscious, the darkness, to discover the deeper truth.

It is here that spiritual birth begins. The labor contractions may crush with tremendous force. A change happens to the seemingly secure and comfortable environment. There is the desire to know the Truth, driven by the hope of relief from pain and the longing to release potential creativity. Recognition of past wounds and false beliefs brings freedom to thought and to love. At this point there is a realization of the ability to survive and to make beneficial choices.

TWO: Drifting in the sea as a vessel without direction.

In this stage there is confusion, fear, loss of prior definition and feeling anchorless. There is the recognition of the force of the All Mighty One and the lack of assumed control over life. The floundering is frightening, for what seemed to be clear and understood at one moment no longer applies in the next. The acknowledgment of not knowing all the answers, nor even one answer, brings relief. It helps to say "I surrender to not knowing."

Here the ego is confronted and assumes a place before the All Mighty One. Yet it is unwilling to surrender to Divine Will completely. Darkness appears everywhere, and the deep pain of being human emerges. This can be correlated with entrance into the birth canal or being lost in the power of a greater force.

THREE: Picking apples from the apple tree. Listening to masters, reading the truth as written by others, observing the truth in action.

Here the hunger to know predominates. There is a seeking of teachings from masters, holy scriptures, teachers of the path, healers, and knowledgeable individuals. Here the ego recognizes alternatives of perception and willingly absorbs the new information. The desire to know God surges up in joy and the heart opens to the truth. The darkness invades to dull the joy. There is confusion as to the reality of the Truth.

The hunger may not allow the full digestion of the experience, so it is wise to sit quietly and contemplate the expanded realities. Collaboration with more experienced people who acknowledge the path of Truth will facilitate movement through this stage.

The mind may question the wisdom that is presented, doubting it, perceiving it as imagination, fantasy or mental and emotional manipulation. Darkness descends to cloud the brilliant moments. There is a rapid movement here from Light to Dark and back again, in an attempt to integrate this wisdom.

FOUR: *Driving the chariot. Mastering the libidinal, the rageful, the willful action, so that this force can be directed toward the All Mighty One.*

Here there are violent swings of passion as the heart opens unto God. The energies of spiritual growth move in full force, dissolving the blocks of fear and congestion, releasing vitality and life. As this force moves, the task is to manage them to allow contemplation to co-exist with love.

Alignment of the ego with Divine Will emerges intermittently and there are moments of the ecstasy of Oneness. We recognize the benefits of spiritual alignment. We see that the way of Divine Will is the path of least resistance.

FIVE: *Entering the stream. Inner courage and conviction respond as a magnet to the flow of the stream. Surrendering to the love of the Whole, bowing to the hand and work of the All Mighty One.*

At this time the ego surrenders momentarily to Divine Will. Life is experienced as a path of synchronicities. All events have meaning and are understood to be useful. Love pervades and there develops a sense of trust in the wisdom of the All Mighty One.

This stage corresponds with entrance into the birth canal, moving with the flow of the muscular contraction, being swept forward through the channel for a purpose that is unknown. If the birth is easy, there is anticipated freedom and release. If there is resistance on the part of the mother or the child, in fear of the expected events to follow, the birth will be difficult and long. It is fear that withholds the flow with the magnetism of the stream. It is in releasing these fears that the birthing process is made smooth.

Time expands and birth and death become brief interludes in the history of the soul. Past lives illuminate present experience and the flow of unity brings peace.

SIX: Burning in the light of the sun. Entering the light in love and trust. Being with no thought in the intensity of God's love.

Contemplation of the Whole in this stage brings awareness of no beginning and no end. There is the courage now to enter directly into the light.

Here the child is born into the light of day, perhaps literally open unto a brilliant light not known to his or her eyes before. There is the Divine moment of emergence in the world of conscious thought and physicality and individuality.

This can occur in life but it is often only experienced at death. In the entering of this Light, karma ignites and burns to endless space. Fear dissolves to lead the way to love and Divinity. Courage is needed at this time to enter into the intensity of this Light. If there is trust in the ability to survive, and the ego can merge with the Divine, this will be a most joyful experience.

Here there is the state of Divine contemplation, of no ego as a separate state. This is experienced intermittently until our passing, for there is need of management of survival on the earth plane. There is great joy in knowing the reality of this joyful place of being. There is a great sense of strength in being able to transcend to reach this holiness.

SEVEN: *Entering the caverns of the tree. Descending into the darkness. Falling, floating in endless space of darkness, pressed into the cavern only to find darkness.*

    Here there is no ego, no sense of I, but awareness of an all-pervasive silken Void. Here there is complete safety in the darkness, despite the initial constriction through the opening of the cavern.

    Pranayama can bring this state spontaneously with forceful breathing of a specific nature. The ego must surrender to the power of the breath, let go of the fear of death, and surrender into the unknown. Here is the experience of ego death while alive, of floating in endless velvet love, egoless, and yet, fully living.

EIGHT: *Flying around the sun. Discovery of the ability to govern direction in space and to enjoy the new-found freedom.*

    At this stage there is motion without karmic magnetic control and the experience of immortality. The ego and Divine Will are united and there is the joyful explosion into freedom beyond karma.

    Here we can choose the state of being and the level of density so desired. The clouds of darkness dissipate to reveal the magnificent land of cosmic colors. Time, past and present, co-exists in harmony and peace. Lightness surrounds, as if walking in space, with no weight of karmic destiny.

NINE: *Birds on a branch. Song issues forth in pure joy of living and of love. Being in balance and in song in daily life.*

    The ego is fulfilled and oriented toward the purpose of assisting to bring joy to the community of Earth. Walking in Light there is the experience of non-attachment and the ecstasy of being in the physical form while experiencing this love. There is realization of the Divine on Earth and of the reality of Cosmic Union. It is not a leaving of the Earth plane but an entrance into the union of the Earth and the Divine.

Here is dedication to the community of God and to the joy that lives within the heart of each fellow soul. This may manifest in the teaching of Divine love or in quietude alone. All who come to know this face will know the light within.

TEN: Naked in the water with a crown. Half in the sun and half in the water, wearing the crown of regency and glory. Clasping the fish, so caught, and to return at will to the water's depths. Balance.

Here is unity of ego, body and soul, of all auric bodies, grounded on Earth, and anchored in the Light at the same time. Male and female energies are balanced and fluid as the sun and the waters flow together. The ego is fulfilled and highly functional, moving with the fluidity of the requirements at hand.

The fish represents the power of the alchemical waters as origin and preservation of life. The crown symbolizes spiritual victory and the highest attainment. The circle expresses the continuity and endlessness of time.

There is complete freedom of movement and immersion in both conscious and unconscious thought simultaneously. Here transmutation is complete—lead is gold, and the circle is completion.

CHAPTER SEVENTEEN
SEPARATION

Guidance leads us to our fears. All fear is based on the illusion of separation. The purpose of guidance is to bring us home, home to our own Self. We choose parents and a childhood environment which manifest in physicality to express issues that we wish to heal. These are locked firmly in place in birth and early childhood and are echoed through the whole life experience. As we look upon the world through transferential glasses, images of darkness appear. The images express misalignment with the Whole and reflect distortion in our vision. The most profound illusion is that we are alone and separate in our beings.

The preconception state is experienced as a state of full soft beauty, often seen in pastel colors. In the choice to take a physical body, we lose consciousness of this state as we reach the denser vibration of physicality. It is to this state that we go at death.

Transmutation requires entering into fear, and our greatest fear is the fear of separation. We have a tendency to avoid and deny fear in hope that it will disappear. Even if the fear is not known consciously, it is held in the consciousness and the cells of the body.

The fear of separation brings us home. Many people born during this century were separated from their mothers at birth. This experience is recorded in memory and governs the nature of many adult relationships. Mothers, and therefore infants, were chemically intoxicated, which intensified the disconnection and sense of alienation. We often have feelings of not belonging in their families, of being misfits. This sense of isolation drives the hunger to know union and motivates the search for spiritual fusion. Unfortunately, some turn to drugs, which creates the experience of blissful union for a period of time, only to leave us plummeting down to our everyday experience of separation.

The fear of separation may manifest as a fear of absorption or engulfment, the other side of the fear of alienation. To be swallowed by the Whole is the same as to be rejected by the Whole. Both appear to lead to annihilation and the ending of existence. Within each there is the longing for union in love and Wholeness.

The beauty of the fear of separation is that its power and pervasiveness call forth the search for God. Through meditation and healing the emotional understanding of the nature of individual separation, fears come to consciousness. However, the source of these fears is the ultimate dilemma of human beings. Because we are individually separate beings in the physical state, we have a longing and an intuitive knowledge of unity. This feeling of aloneness creates a deep sense of isolation.

Message:

As we delve deep, deep within the cavern of the heart, we enter into the fear that lurks in the darkness. Just as the descent to the womb of the cavern occurs in sequential steps, so the journey unto the ultimate human fear, the fear of separation, proceeds in stages. Being the fear held deepest in the human heart, a sudden descent would so traumatize the heart that death would naturally occur. Thus, the journey is a gradual one, undertaken with the knowledge of the Whole. It is this knowledge of the truth—that all are One—that leads, nourishes, and provides the bed of rest for the spiritual traveler.

As the fear of separation is the fear of aloneness, it is beneficial to enter into these depths with hands holding one another. For as the darkness looms, the human hands provide the warmth of life embodied and shared. It is not meant for us to be alone, but rather for us to come to know that we are not alone. Our fellow travelers bring openings in the shared experience of humanness. The presence of the bodies, warm and living love, give vital nourishment along the path to God.

The bliss so sought exists within the descent, facilitated by the loving hands of transformation. Know that You and I and all are One and each needed by the other.

WALL OF SEPARATION EXERCISE:

1. Sit quietly, eyes closed. Breathe deeply and imagine a wall before you. Notice the color, shape, length, width and height. What consistency is it? Notice how it feels to sit before this wall.

2. Speak to the wall and ask it why is there. Listen to the answer. Notice how you feel. Ask what purpose it serves you. Listen to the answer.

3. Now become the wall. Feel the depth and width and height. Feel the consistency. Feel the weight and strength. Say to yourself, "I am the wall and it feels...." Be aware of this feeling.

4. Become yourself again and notice how it feels to be with the wall. Say something to the wall and let it answer.

5. Imagine an opening in the wall. Enter this opening. What do you see on the other side? What does it feel like to be there? Who or what is there? Look back at the wall and notice if it looks different than before. How do you feel about this difference?

6. You have entered your fear and have gone beyond it. How does it feel? If you did not enter your fear of separation, do this exercise again. Do this knowing that the wall represents your fear and that moving to the other side of the wall represents your successful overcoming of fear.

CHAPTER EIGHTEEN
PAIN

In All There Is, there is no pain, no suffering, no holes. There is only Light. And yet, as human forms, we cry, we squirm, we hide our heads in shame and helplessness. We fire the heat of anger against one another for self-protection and the illusion of security. We ourselves create the pain to know our reality, to know the depth and breadth of God and all its creation.

We bring with us the memory of prior pain, so known for eons in the physical form. It is, of course, the pain of separation that is most deeply felt, so profoundly known, that we identify it as us.

Let not the emptiness dissuade you from your task of knowing All There Is. Let not the pain reside within your heart. Be your governor. Allow the fear to coexist within the womb of human form, to coddle you, to cradle you, to sing to you a lullaby of love.

Pain sinks into the very essence of your being to be as a beacon, shining light upon the deep waters of the soul. Yes, you must all feel the pain of living in a human body, of birth and death and injury. Yes, you must all agonize in love, in loss, in fear of loving. For this is the reason for the pain.

Through unveiling of the Shadow pains, the light of truth shines through. Through revealing all the darkened shapes, unlit within the soul, the light begins to grow. Know, without darkness, Truth beams not, for the other side is absent. Though all is one and one is all, the Light reflects the Dark. Seek comfort in the darkened shapes that lurk within the wretches of your heart. Traverse the murky caverns of your mind, the crevices and cracks. Know the Light is there, in truth, and know in all the haunted fears, in all the vicious arguments, in all the tribunes held, the sacrifices made to God are all within these walls. To die and be reborn. To merge beneath the conscious mind. To squalor in the muddened pit, and yet, once again, to rise to find the Light.

In the mind thought forms develop to contain and define the pain. These may consist of: "I am alone." "I am unworthy of divine union." "I am bad." "I will be left." "I destroy." "I am ineffectual." "I create pain." These beliefs permeate our consciousness, and living involves creating situations that enact or deny these beliefs.

On the emotional level pain is separation anxiety in its many forms. It begins prior to conception and continues, unless there is healing around death. At these two entrance points, birth and death, pain is accentuated because of the force of pressure of the magnetism—first from the Whole to incarnation and then from the incarnation to the Whole.

Our nervous systems feel physical pain as a reminder of a dysfunction in the body. When the body functions as an integrated whole, there is no pain. Pain can be a welcome warning of the need to reorient the Self to heal the pain. For some, however, there can be an attachment to the pain for reasons of drawing attention and love when there seems no other way. In the midst of pain there is the awareness of being real, of something to hold on to, when all else seems to fade away. It is for these reasons that there may develop an addiction to the state of pain, which may resist the healing.

In transforming pain, all levels of being — spiritual, mental, emotional, and physical — must be addressed. Basic to the transformation is the enlightening of the Shadow, the disclaimed, projected, undigested aspects of the Self. Spiritual practice can help in the realignment and lightening of these darkened parts, fostering a belief in the power to change and trust in love.

As the wall melts that separates us from the Whole, conscious awareness of the reality of the All Mighty One grows. There is, however, pain in the remembering, for before the fusion there is resistance to knowing the reality of the beliefs that keep us separate. It can be likened to the healing of a broken bone where there is the break, the trauma, and then the pain of realignment and the resistance of the muscles in adjusting to the change.

If there is severe physical trauma, which cannot in this time and place be readjusted for alignment, the experience of the pain can change in healing. The sensation of pain may remain the same but the interpretation of the pain will be different. If we change our belief from "I am bad because I have pain," to "I am beautiful and whole, and my pain is part of me," then the experience of the pain itself, as well as our understanding of ourselves, will change.

Physical or sexual abuse in childhood is common in the experience of those suffering from chronic pain. This abuse intensifies the identification of self with pain. It is necessary to heal these emotional wounds as well as the physical wounds in order to realign the self with positive living.

Pain is a reminder of human vulnerability, often stigmatized due to projected fears of this vulnerability. It is important to remember that no one chooses pain consciously and that all pain is real. Pain is a signal calling us home to our true nature. In our stubbornness, this may be the only way we can turn to the All Mighty One. If pain leads us here, then it can be with profound gratitude that we may thank the pain for this direction.

Pain is our teacher, our karmic staff, our guide to God. It is strange that pain can lead us to peace, and yet it does. In the most painful moments, courage and faith move us to knowledge of the Truth. Those who cannot find this Truth at the chosen time die and are born again later to move closer to God. We are born again and again until we reach this Truth in our hearts and in our souls, in our total beings. This will not be our last chance. However, this lifetime offers us the chance to transmute completely, and it is this that drives us in so many convoluted motions to find this Truth. So many obstacles, like wrong turns in a maze, come before us. We make many attempts to find this Truth through sex, alcohol, drugs, food, gambling, swimming, jogging, mountain-climbing, scuba-diving, writing, dancing, singing, horseback riding. The list goes on and on. All activities, all life, is directed, perhaps unknowingly, toward the search to know the Truth of union with All.

When we feel joy, exhilaration, peace, ecstasy of arousal, we feel the Truth of union, which drives us to know more. Beneath the pleasure there is the quiet, or not so quiet, sense of desperation. We believe that we will not find the Truth, that it may be gone forever. There is the voice within that doubts, that fears, that speaks of self-delusion and doubts our union with the Divine. There may be a misguided belief that it is better to be alone and separate and away from the entanglement and pain of relationship and commitment to another.

Our world is founded on the belief in pain and fear as the mechanism of social control. The question arises, "How would we control human behavior without pain? How could power be managed without the weaker and the stronger?" There are the more personal questions: "Who am I without pain? Why would someone love me without my pain, without my suffering, without my weakness?" The answer lies within. For as self-love arises to melt the pain and fear, greed, anger, and jealousy recede. There is the All and within the All there is enough for all. Each person has their unique and precious combination of abilities, life circumstances, desires, and hopes. All are intertwined in the fantastic web of Love.

As pain lessens, change takes place within the individual and the group. Values shift to loving and creativity. Disengagement in destructive situations ensues. There may be fear in anticipation of this disengagement, for this change appears to be the cause of the pain of separation. However, as love grows, pain is transformed into faith and trust, and the changes occur naturally and without the anticipated pain. Those who are left in the darkness do feel pain until they, too, open unto the Light.

Pain is the longing to unite with the Self—to be whole and full and not alone. This is the longing to know that which we have known and have now forgotten, and to come home to ourselves and our glory. Within pain is tremendous potential for growth.

When pain is held in, there is tension and the augmentation of fear. Defenses can build up to prevent the awareness of this pain and more tension develops. This tension is likely to be felt physically as well as emotionally. Muscles become chronically tense, and therefore, both the muscles and the bones they support are more prone to injury. The fear and sense of chaos, helplessness, and powerlessness demand most of our survival and attentive energy. This is potent, unharnessed, unfocused energy. In fact, within the pain is life energy itself. Deep within the well of pain is the source of inner strength, love, joy, and peace. Within pain there is God. It is not to dwell on pain, but rather to move into it to find the Truth.

Message:

Hail forth, my beloved warriors. Let it be known, you each have the exact amount of courage needed to fit the tasks at hand. Let it be known, there are councils guiding you as you traverse the shadowed light. For you have chosen life, and life holds you firm until you choose to die. The passion that you seek, indeed, lies within the molten Shadow of your heart. With glasses on, you enter as a miner to a mine. Each moment risking death and choosing life. The glasses provide a screen to filter that which you experience within the darkened depths,. Thus, protected, as you wish, your journey does begin.

Pain holds Truth, unconscious wishes, fears, fantasies, knowledge. Though so often felt as an enemy, pain is a friend, an ally, a monitor. It is a reminder of the misalignment and the distortion of the Truth. This does not mean that pain is not real, for indeed it is, both emotionally and physically. Not one of you would choose to feel the pain in either way, and so, as a messenger it comes to remind you of the trouble in the heart.

Lift now to meet the All Mighty One. Allow us to welcome and caress you in the fullness of our being. Belong to that which you truly are, and know that You and I and All That Is are One.

Pain is forgetting who you are, forgetting union with the All Mighty One, and the search is to remember once again the truth of who you are.

CHAPTER NINETEEN
TRANSITIONS

How distant and mysterious an idea looks from above, and yet how rich and fanciful it looks once we are immersed within the vision. So it is with transitions.

Observed from distance, a transition appears fearful, dark and dreary, infiltrated with pain and horror. And yet, once submerged within the transitional action, life itself picks up the force. What was anticipated as painful, empty, unknown, and uncaring is filled with teeming forms of life and the full force of beingness.

Each moment is in itself a transition, as is each hour, day and year. Time moves through space creating movement, one force upon another. Indeed, time only speaks to those who stand apart from it.

What is time? Time knows All in All. In All there is no time. And yet, from the limited perspective of the human self, time cries in anguish at the deepening gaps and gullies of the ravages of time. Time knows remorse and guilt and loss. Time knows of love. Time knows of order—order in the beingness of all. Should you decide to enter time, you would find a sound of entrance pass, a sound so filled with holiness to move you beyond the veil of limitations, so pressed upon by time.

Yes, it is possible to enter time through alteration of your consciousness. Sit quietly and breathe. Go deep within. Allow the breath to move in the rhythms of the flow, so moved to enter spatial non-dimensional wholeness. Know within this space there is no time, and yet there is full presence. Know this is the presence that occurs after death, the bliss, the joy of All.

In the moments past death, the moment of the transition is never to be known again in life, and yet will be forever there in spirit. Human life transitions are marked to alter consciousness, to open unto Light. But darkness looms heavy forward of the Light. Anticipation of death creates resistance to the flow. Time sharpens its hold, teeth clench, and fear builds in hope of that which is inevitable might not be so. Eyes close to block the future birth. The opportunity is present to open unto the dawning Light, to loosen the past, and yet we may instinctively hold tight and fight the Light out of our fear of death.

Choose carefully a friend to guide and hold you through the tunnel of the unknown. Open you heart in love and faith at last. It is this opening that will bring you to the Light. The love so known will embrace the change, love so deeply felt that brings you unto God. Deep within, the stirrings of love are reversed, so we love our own hearts as we feel God's love.

Transitions mark the passage of time in human consciousness. They call forth and challenge all that is known to be truth, exposing limitations of concepts and beliefs, catapulting creation beyond the present form. These may take place in full consciousness and choice. Or, as often happens, transitions may be as the surge of ocean waves, forced up and out, to be pulled crashing down by the full force of magnetism. Transitions evolve in a multitude of forms, each perfectly designed to reach the further step of growth, so situated within the human destiny.

Birth

From preconception to entrance into the world of material form, the first and foremost transition sings the glory that is life. The movement through the cervix into the birth canal is referred to as "transition," which is a time of immense force, the pressure of a thousand waves. From a state of quiet and repose, thrust into individuality, the child emerges to enact the blueprint of this life.

As childhood unfolds, the auric layers gather form, guided by the blueprint of our lives which governs physical, emotional, mental and spiritual development. From birth until death, the planetary influences impact upon our auric fields, intricately woven into the blueprint that is perfectly designed. Each and every experience is recorded, influencing our growth.

The physical body is the crystallization of the entire energy field, continually vibrating in interpenetration with the outer environment. The Self grows to know itself, from union with the mother's form to the experience of separateness and difference. The energies flow to the areas of least resistance, seeking to avoid blocks within the field where experiences of pain and fear distort true reality. There is continuous motion toward Wholeness, and the blocks contain the very energies required to free our full potential. Thus, interaction with the environment creates the obstacles needed to break though in order to free these energies. We spend much of our early life, and often the whole life, creating and struggling with the repetitive patterns set up by these blocks.

Adolescence

Adolescence is the next profound transition—from childhood to adulthood. As in birth, there is descent into darkness and the unknown, into chaos and confusion, into self-doubt and delusion about the Self. The chakras come into full form, and adult sensuality and sexuality arise to heighten hunger for life's meaning. There is a movement away from family constraints and expectations into freedom and self-liberation. As with all transitions, some flow through smoothly without turmoil, while others enter profound pain and loneliness. Still others stagnate, regress or become petrified. If we move back in consciousness to birth and adolescence, we will discover the patterns we set in motion as to the transitional flow. We can alter these patterns with love and higher consciousness as freedom from the holds of karma release the unconscious compulsion toward repetition.

Midlife

The midlife transition often brings unexpected pain as life no longer means the same. The first half of life involves movement toward achievement and acquisition, family, home, work, external security and success. In the transition to the second half of life there is a searching for the meaning of the life, the purpose for which one incarnated, the fear of mortality, and the apparent emptiness ahead. We ask "Will I move on? Am I destined to repeat forever the self-destructive patterns I have created? Why am I here? Who am I without my outer definitions?"

We need an act of Will to choose to step into the darkness, to step into the void of the unknown sense of Self, and produce that which is desired. Ahead there seems only aloneness and more pain, and yet we have an inner drive pushing us into the darkness in hope of self-discovery.

Resistance to this inner movement may manifest as sleep disturbance, anxiety, unusual fears, attraction to younger lovers, intense interest or disinterest in sensuality, and grief. Often we have a vague sense of hopelessness and fear, accompanied by doubts, dryness, sterility, sexual dysfunction, and emptiness.

The second half of life can be referred to as a ripening, a time of fruition and completion of personal tasks, of moving into life purpose. Our life purpose is intimately intertwined with the soul, for each soul holds our blueprint that travels through lifetimes. Transmutation is the process that unites the personality to the soul's energies to enable the soul's task to be enacted upon the earth. The time of midlife is crucial to bringing about this process. That which has been lived before creates this meeting of personality and soul through transmutative processes. This is the reason that transmutative processes are heightened in midlife. In transmutation there is the heightening of rebirthing, creating profound change in consciousness and resulting movements through life. There is a shift in perception that may alter personal attraction, the nature of desire, attitudes toward self and others, interests in activities and work, which may confuse those within close proximity.

The midlife transition is one of profound significance, for it highlights all that which has come before and all that will follow. Time alters its course. Children disappear and reappear as adults.

The wind blows heavily through the trees as the force of nature makes itself known. There is a deep ache in the heart as we recognize the transitory nature of life on earth. The soil is fertile and fallow, having enlivened, nurtured, and reabsorbed the products of the first half of life. This is a time of rest and re-evaluation, while waiting for the next planting.

Message:

In contemplation of midlife, rest now. Rest in peace, for there is much ahead, as there is much behind. Know that now is a time of contemplation and being. Know and accept the impatience that wells up inside you, fervently calling you to go forth. Trust and speak deeply from the heart. Full knowledge of the former living emerges. You see the decisions made, the dreams achieved and crushed, the twists and turns that you created, all bringing you to this moment.

Allow full recognition of the choices made from incarnation to the very present. Know that you have chosen to sit in contemplation with me at this moment to explore the meaning of our communication. Know you have asked to know, and this is the very essence of the mid-life transition. To ask is to be heard and responded to. To ask of pain and sorrow, of disappointment and loss, of joy, of peace, of your heart's fulfillment. All these questions come tumbling out, flooding the mind, opening the heart.

There is time. For what is time but being, and being is always becoming. A moment can be minute or an eon, depending on the soul's connection to the Whole. All this consciousness of time expands and grows. Allow the present to come forth fully unto the Light. Allow the Truth in all its many colors to emerge from your deepest depths, to free you from the bonds of holding on. Let fly the glory of all that you are, and let us guide you forever more, holding you in our arms as we walk together amidst the glory that is All.

Know this journey is the most important you will undertake, to rediscover that which you truly are, to remember the aspects of the Whole, long forgotten and discarded. Know there is the deepest peace within the human heart. This peace lies before you and seeks expression through the love that envelops all. Know that in your seeking, you will find.

Twilight

The next major transition is when we move from midlife into twilight. This can be a most enjoyable time if the transition is made from identification with the physical body to a contemplative, spiritual focus on beingness that is separate from physicality. This transition is often darkened in Western culture by emphasis on appearance, youth and the speed of living. Twilight can be a time of peace and reflection, remembering and reuniting all aspects of this lifetime. It is wise to express this in a creative form, such as writing, speaking, painting or sculpture, sharing and inspiring expanded consciousness in others.

In twilight, we move from our tendency to "will" life events to happen, and instead trust in the hand of the All Mighty One. For a man it is predominately to release identification with physical strength and for a woman to release identification with physical beauty. Allow the light of the soul to shine forth in your eyes, showering upon those that witness this light strength and beauty far beyond physicality.

Due to the focus inward and the lightening of the aura, the concentration on material or mental tasks may lessen. This is a natural movement toward the Light, a pull toward that from where you came. If it is resisted, there is frustration and discomfort due to confusion and fear. Relax into the natural change and allow your consciousness to expand to thoughts of the All Mighty One and the beyond. Fill your heart with love for that which you have created. Forgive all misconceptions and misjudgments. Release the fear, for the process is inevitable at this time. There will come a time of immortality when choice as to length of life is conscious. However, this time has not yet come. Thus, physical death comes without choice in conscious thought.

The body until twilight has functioned as an efficient and effective mechanism, carrying you through the material world, responding to the Will. Now it is in need of special nourishment and care in the form of gentle, flowing exercise and a diet that your body will ask for, should you listen.

Death

In approaching the end of this life know that Darkness is but the absence of Light. It is the Darkness of the fear of separation that is beneath it all. The association with Darkness is unknowing, as a ship lost in a winded storm with no guidance toward the Light.

In truth, it is the fear of the wayward state without indicators, chaos, that is the fear of death. For in this state there seems no guiding Light, but in truth the Light is united with love. So we imagine that in the state of anticipated darkness, there is no love. If we choose to live to know love in the physical, sensual form, and therefore to know the All Mighty One in a conscious, physical state of being, we proceed to opening to the Truth and to the One, and we know the One in full physical consciousness. To know God—this is the true reason for living in the physical form.

At death there is the return to the effervescent light of love. This transition of death is one of enfolding the elements and senses back into non-material form, becoming detached from the physical body. We can be drawn back to the psyches of those left on the Earth plane, particularly if there are unresolved, unforgiven situations left behind. This creates a struggle and a more difficult transition. Releasing blocks developed through our lifetime facilitates the passage through this time. Healing on all levels is highly recommended to proceed easily in this movement to the Light.

In the transition, there is separation from the physical realm. Fear of separation and pain arise to be confronted once again. The tunnel so often described is the projected form of transition, parallel to birth. Healing the trauma of being born assists profoundly in the movement through the death tunnel, for it is at these two junctures that the magnetic shift is strongest.

Less monumental transitions evolve throughout a lifetime, such as a job change, the birth of a child, the death of a loved one, marriage, the ending of a marriage, or a geographical move.

Message:

Surrender now the Will to Divine Will. Listen to your intuition. Open to love and the expression of love. In twilight there is a need for friendships, for sharing wisdom gained, for reflection on the Truth and perception of life events. There is also a need for a realistic assessment of material security for you and those you love, for often there arise the fears of early life, which may distort the Truth. Release to the All Mighty One your care and your deep love. Know that all is in perfection, and it is only your fear that may prevent you from the glory of the twilight.

The entrance into the darkened Void to be reborn continues through any transition, from a moment of time to birth and to death. Know these transitions come for opening to love and for releasing the blocks that obstruct movement toward the Light. Hold high your sword of truth, and enter, beloved ones, into the Holy Void.

CHAPTER TWENTY
BIRTH AND THE AKASHIC RECORDS

There is tremendous creative force instilled in the process of each and every birth, animal and human alike. The energetic transformation compacts into life multitudinous frequencies of energy, from the densest of bone and hoof to the lightest of thought and consciousness. Birth brings forth these energetic forms in a living organism, which functions simultaneously on all these levels. Birth is indeed the birth of a magnificent creation.

The *Akashic Records* present in multidimensional form the presence of energetic transformation. These records are known to the human mind in forms that can be understood, but in essence they exist in pure energetic form. At birth there is a spark of knowing that emerges from the energetic flow to be condensed and bonded with the physical energetic formations. The combustion of this union imprints upon the creation a blueprint of the life. This blueprint exists in potential form and at different levels of frequency within each life being. Karma is contained within this blueprint. Existing simultaneously is Free Will, which energizes the blueprint and provides the potential for creative growth. The blueprint holds the pattern while Free Will expresses the potential for expansion of consciousness, (i.e., the response to the patterned events).

The imprint at birth contains the influences of multidimensional forces, including planetary and atmospheric magnetism, the auric body of the mother, and to a lesser degree the auric body of the father and other beings around the mother. It also contains influences of nutrients absorbed, the genetic coding from mother and father, and the physical and emotional conditions at the time of birth.

The emergence of birth presents to the Earth's electromagnetic field a glorious creation, imprinted with learning and with the potential for contribution. Know that you and all other life forms are in essence pure creation.

CHAPTER TWENTY ONE
LIFE ENERGY

At the moment of conception the Light within the cells are ignited at the point of combustion through all the elements: earth, fire, water, air, and ether. When the fire is lit, the earth quivers, the air sings, the water flows, the ether waves. It is true for all beings. The difference between animate and inanimate beings is the vibratory rate within and between the elements.

The elements contain vital properties of a very subtle nature. Conception occurs as a shift in the magnetic force in the receptor and the sensor (the ovum and the sperm). The elements instantaneously create combustion as the two strike each other. In childhood the elements continue forming and reforming, according to the blueprints of the receptor and sensor, blossoming and flowing in all directions.

When the elements reach a stabilizing point around mid-life, the combustive energy begins to diminish, as if resources are gradually being consumed. The combustive energy can be regenerated through exercise, diet, joy, and love, most of all. Love creates increased movement, exciting interaction between the elements. Like sunshine, love warms and soothes. Love creates the matrix in which the elements vibrate, vibrating itself also. So as love intensifies, the vibration of the matrix intensifies. This is true for all living beings, the loved and the lover both.

In mid-life there is a turning inward to seek love, and a fear of finding no love there may arise. When there is fear, the vibratory level decreases. Disorder and disease can occur as the elements are not interacting at a level that maintains vitality. The vibrational order becomes chaotic and weakened. Such chaos is life-threatening, and thus, alteration in life style is necessary so that the elements can resume a vital vibratory rate. The door to change opens unto love.

For all aspects of mothering, knowledge of the interpenetrating forces within and without can assist the mother in providing a nourishing and enlightened entrance to the Earth plane for the child. As the mother heals her pain and her fear, so too will she heal her child's. Love and conflict of the mother and father is incorporated into the child's consciousness and is present within the child from the moment of conception until death. The blocks in the mother's aura directly influence the development of the fetus, so transformation prior to conception and during gestation can be helpful in easing this transition for both mother and child. A hearty welcome to the being who has come to give and to transform is a profound blessing for all within the family sphere, for the birth is one for all.

CHAPTER TWENTY TWO
POWER

In love there is no power but the power of God, the All Mighty One.

M<small>essage:</small>

Seek not to calm the power. Speak not of heaviness and fallacy. Sever not the bonds to your own Self. Dig not to farrow out the Darkness, the dungeon demons. Cry not for lost souls, so hung in shame. Dampen not the fire that burns within your heart, the fire that sears scratches upon your breast. Divine, instead, the hallowed halls of Truth.

If you go in faith and holiness, the paths will open unto thee to show the way. If you seek authority to govern from the position of above, slow will be your progress unto God. Know that which you say you fear is naught but that which you so dearly love. You of many minds behold the promise so bestowed upon your upheld hands. That promise soars to hold the sky and melt the Earth. That promise so bestowed in love seeks only to be known. That promise whispered to your soul now moves to self-expression. That promise is WHO YOU ARE. To heal your fears will bring you close to God.

No fear is here with God. The fear of power is naught but the Shadow's depths projected onto stone. Sink deeply, deeply into love. Release the Shadow's hold. Know that which you have known forever and forgotten once again. Remember, hallowed be the Lord.

Remember who you are. Know the Shadow holds the darkness of many lifetimes. Moments of enlightening will erase the scars of old. In faith, move perfectly in balance, unfolding step by step. Know impatience is but fear of the unknown. Know that which lies within is God, perfect unto itself. Hunger and impatience fade. As dusk to dawn, the glorious Light shines forth in faith and love. The power that you seek is but the love of God transformed. So find its place within your heart in harmony and peace.

Know that each of us, as each of you, holds talents from afar. Know each of us, as each of you, has expertise to share. Power, as you call it, is necessary to fulfill the tasks ahead.

Swiftly, as the runner flees torrential rains, the sun emerges, and brilliant light appears upon the same situation from which he ran. Thus, know that from which you run can become anew within a flash of light. The voice of God calls forth in many tones and you shall hear that which you listen for.

CHAPTER TWENTY THREE
APPROACHING 2012

As we approach 2012, now is a time where attention is turning to developing power *within* rather than power *over*. As the Shadow aspects are lightened and love is released, the need to exert power over another for personal and/or public safety will diminish. In mid-life there is a natural turning inward to find power within, although there is often the fear of finding emptiness or soft mush.

Gestalt and Jungian psychotherapy are effective in reinforcing the awareness of inner potency and substance as they magnify the Shadow aspects which withhold this energy. By illuminating the Shadow projections through work with dream images or the empty chair (where imagined significant others and parts of the self are placed to be known and integrated into the psyche), you can claim the dark, hidden areas of the personality and bring them into the Light.

Message:

Potency is but another name for being. Being, as in the presence of one enlightened from within. The degree of opening to Light and love, forthcoming, will be the magnitude of one's potency. Know darkness fades as Light emerges from the vessel held within the soul. This vessel contains the nectar of love, seeping through the membranes of the soul. As this essence penetrates the outer layers, the heart opens unto sweetness unknown until this time.

Know it is time now to employ this essence unto the world without and to enact the inner world upon the Earth. No longer can you sit upon the mountain ledge in reverie and contemplation. You must now initiate the transmutation forthcoming among the peoples of the Earth. Each of you, as each of us, holds potency for enactment and for direction in the construction of the forthcoming structures upon the Earth. We have spoken of the desire within the soul to manifest a form of love that will bring solace, healing, and love to those of you less oriented toward leadership.

In each of you there is a unique and precious formula of presentation that holds the potentiality of creation in the world of form. For it is a time of manifesting creative forms of healing. There is always choice, however, in the degree and magnitude of such a form, as if plans are drawn and workers await instructions. Know also all ingredients required for the manifestation of creation lie within. It is Free Will that guides you through each twist and turn upon the path.

Potency, as being, moves unto the magnetic force within. As magnetism magnifies, the vibrational level alters. We as beings magnify our full essence, and when there is a being on the Earth plane vibrating at the self same rate, we join, and there can be a guided mercurial transmutative meeting. Choice is of the essence here, and potency within is the guiding force. Most essential also is the degree of psychological well-being, which provides the fertile earth upon which the meeting grows. Know there is no separation within the process, and in describing it here, there has been a contradiction suggested that is not so in the true form. And that is, that you and I are different. For in truth, we are one with All and the All Mighty One, and in the meeting, this totality is known. We are the chariots of fire for you to traverse the molten sky in love. We are of you, and serve you as the love of God.

Ask and you shall be heard, my love. Know that I am here within and am present with all you seek to know. Know in truth that we all are one vibration of form and essences in the glorious manifestation of love. Seek to find your own Self, and you will find that potency emerges naturally and without effort.

POWER EXERCISE:

1. Place your hands and fingers in the mudra "OM." This is created by the right hand resting just below the third chakra (solar plexus), cupped slightly, open and parallel to the ground and the left hand with first finger touching thumb in a circle in front of

the solar plexus with the wrist turned so that the palm faces forward. Feel the circle of energy created between your hands. Allow this energy to penetrate your third chakra.

2. Imagine a situation of power you are engaged in now in your life. Notice the details and the feelings of being powerful and in your own power.

3. Now move back in time in your life to another time of engaging in power. Notice the details and the feelings. What is the truth here?

4. Continue back further in your life to another time of engaging in power. Notice the details and the feelings. What is the truth at this time?

5. You may continue to do this as many times as you wish, but always return to the present situation, and ask, "What is the truth?" This is so that you can discover power conditioning, and the truth in each situation, which is probably clouded by projections and fear. By bringing the awareness back to the present, you can melt the obstacles.

6. This mudra (the physical act) can be used whenever there is fear of being overpowered or of overpowering. The difficulty will be in remembering to use it at these times. It is helpful to know that this is an action that will have a direct effect on your third chakra, and can help you whenever you choose to use it.

CHAPTER TWENTY FOUR
ADDICTIONS

The fear of separation manifests in habitual form, limiting creative instigation and expression of the potential held within the blueprint of the soul. When habitual pathways solidify in frozen fear, the habits magnify to become addictions.

So often within the addiction lies a darkened cavern immersed in pain. This darkness is of the pain of disillusionment and betrayal, the belief that which one had incarnated to become could not and would not be. As within the cold underground recesses of the cavern, darkened energy resides deep within the addicted person's solar plexus, waiting to be released unto the heart. The patterns, compelled by the thought forms, vibrate in rigid motion, limiting, frightening, saddening the one who carries them.

Enfolding one upon another, these addictions mushroom to become a style of living, attracting those who share this patterning. Thus, we build shell around the darkened hole to block that Light in hope of reducing pain. However, the dark energy continues to engulf, to masticate, to ulcerate the inner state of being. Illusion upon illusion disturbs the fragile equilibrium.

In this state of habit and addiction, creativity is stymied in the rigid bonds of fear. Attempts to reach the heart to find the Light are obstructed by the fear itself. We believe that survival itself depends on these behaviors and on the sustenance outside the self to provide the comfort from the pain. This may be an addiction to drugs, to alcohol, to pain, to fear, to love, to sex, to material pleasures, to another being, or all of the above.

It is the fear of separation from the Whole that originates the darkness. This fear must be dissolved in love for the addiction to release its hold. For the nature of the habit may change, but the habitual patterning may stay the same, and thus, one habit may become another easily. It is the healing of this pain, the pain of separation from the Whole, that will melt way the patterning.

When we turn our intention to love of the All Mighty One, all need for covering of the darkness dissipates. The warmth of the meeting of the heart with God releases the frozen fear. Whenever there is constriction in the energy field in the physical, emotional, or spiritual layers, we experience a signal of the rigidifying. These signals warn of divergence from the path of one's own heart, thus providing guidance for the flow of love and consciousness.

It is in alignment of the Will with the Divine Will that the addiction is transmuted into the fiery love of the All Mighty One. The creativity held within the darkness flows free and unobstructed. Initiative abounds in the joy of freedom and in the acceptance of the true reality. The fear of separation no longer binds the individual's contribution to the self, the other, and the Earth.

Lift unto the love of your own Self and know that within the pain there is the glory of the One. Allow the opening, the venture into love, and you shall find the glory that is you.

The tendency to resist change limits freedom of response and responsibility. As willingness to remain in the present moment increases, the ability to respond spontaneously and in accordance with Divine Will increases in exact proportion. Here we experience the flow of life as the glory of the Whole, as the Perfection of All That Is. Faith in oneself and the All Mighty One grows and reinforces this state of responsibility.

Entering with the present may create a fear of losing control, of hurting oneself or another, or being hurt by another. Ultimately, there is the fear of annihilation and total destruction. We fear we will regress to the primitive state of being without limits and without conscience.

Growth toward true freedom, however, is not regression to this primitive state, but rather, ascension to a state of consciousness that exists synchronistically between inner and outer control. Moving vitally through the present, the choices we make in each moment reflect alignment with Divine Will and provide safety and security in the awareness of the beneficent All Mighty One. Faith alone will not offer this state. There must be action and consciousness of the present state reflected in each motion.

It is the interpretation of events that breaks the hold of the addiction or habit. If there is the belief that destruction will take place if the addictive behavior does not occur, there will be strong resistance to releasing this pattern. If, however, there is an opening to letting go of the pattern, and a willingness to learn new paths to freedom, the love and understanding of the self and others can alter this belief, and in turn, alter the behavior. As this opening occurs, it is as if the edges of the human shape soften and become rounder and more flexible, allowing for fluidity and freedom of movement. Where there previously was a narrow range of response, now there is a relaxed, multi-fold state from which to respond. The fear of destruction that infiltrated all action fades away, allowing love and trust to flow into the spaces that were once rigid and constricted.

Message:

As the bear hibernates in the wintry cold, so, too the vicissitudes of frozen motion instill immobilizing fortresses. As the seagull glides above the salty waters in search of motion below to indicate a fish, so, too, does the mind glide above the active form. When the object of intention is revealed, action brings satisfaction of the desire. The attraction draws as a magnet, compelling action toward the source. There is, however, eternal choice, despite the mind's intention.

As humans, there exists the tendency toward habitual action, regularizing these desires, so that the choice is forgotten for the sake of the trodden path.

Habitual action creates alteration in the mind to correspond to that which has been constructed. This alteration in the mind then contrives to bring to fore the illusion that the action is the only manner in which the task be done. The hardening of this illusion evolves to become situated as belief, which then governs subsequent behavior.

In the silent bell there is no sound, no movement to alter the stillness of the air. So, too, in inaction there is no outer change, no interaction with another force. Change is feared, for so often change has brought destruction in its path. However, as the bell can ring forth in magnificent resonance when struck, so, too, can the force of change within the human life bring joy and resonant beauty. The fear of separation from the Whole rigidifies the pathways of the mind and prevents the venturing into new landscapes of human being.

ADDICTIVE BEHAVIOR EXERCISE:

1. Sit quietly, eyes closed. Open your chakras and breathe in the breath of the All Mighty One. Consider a pattern of behavior you have that you believe you cannot change. This can be as simple as brushing your teeth at a certain time, drinking several glasses of wine each night, saying "Have a good day," whenever you say good-bye, or more destructive behavior patterns.

2. Recall precisely the events, the feelings, the beliefs, the actions, that preceded this pattern. Repeat them in your imagination, completing the pattern.

3. Repeat them again and stop before the actual pattern begins. Notice what you are experiencing in your physical, emotional and mental bodies. Take deep breaths and imagine a red-orange line of energy emerging from the base of your spine down to the center of the earth and back again. Allow this red-orange energy or light to envelop you as a flame in love.

4. Place a picture of you performing this addictive behavior in your heart, as if you were watching it on a television screen. Allow this red-orange light to bum through this scene as a flame, melting, evaporating this scene. Feel the heat of the sacred fire burning. Let the flames move up to your throat, and out your throat, up around your head, down your back , to the base of the spine, and up the front to the heart.

5. You are now encircled in the red-hot flame of the All Mighty One. Now let the flame enter into the heart and fill the entire cavity of the physical body. If there is any residue of the scene in your heart, allow the flames to encompass these remnants.

6. Take several deep breaths. Bring the flame back to the base of your spine. Allow it to remain there to be ignited whenever you wish. Repeat this exercise twice a day for a week, then at least once a week until there is no desire for the addictive behavior.

CHAPTER TWENTY FIVE
DREAMS

Dreams are condensations of energy held in the emotional layer of the aura; however, they simultaneously reflect energies of all other bodies. Their light is of a teacher, illuminating consciousness of the impact of influences from the inner and outer environment. Theirs is the means of expression available to all living beings, which allows integration of experience.

As an artist mixes colors on his palette in order to discover the color that vibrates with his inner sight, so, too, do dreams combine elements of consciousness in order to present to the conscious consciousness a reflection of daily experience. They come as teachers to express the nature of the flow of energies within the aura. When there is an obstruction in the auric field, the dream will illustrate the substance of the obstruction and offer an opening for healing.

Message:

As impurities fly when metal is heated, so, too, do dreams express the residue of the velocity of auric life. They often precede transformation in the lower bodies in order that the substance of the healing can be recorded in the denser energy forms of consciousness. The recognition of this need for transformation may be accepted and acted upon, releasing darkness, or it may be embedded once again to combine with other auric energies and return in another form of dream.

Dreams offer healing potentialities to prevent the contracted energies from reaching the densest form of congestion, that of the physical body. Thus, recognition of the darkness held within the dreams will allow release of negativity and avoidance of physical malfunction.

Dreams reflect openings in all bodies of the aura. Although there may be need for transformation in the physical, emotional, or mental bodies, a dream may reflect opening to the astral or fourth level. Such a dream could be the manifestation of a spiritual teacher, astral beings, or the presence of pure Light. If the message is absorbed by the dreamer as the energy of healing, there will a direct effect upon the three denser bodies. If, however, the darkness of fear prevents this absorption, there will be no physical effect. Surrounding all of you are most beneficent beings who wish to serve and heal you. It is you who must welcome them into your hearts and into your dreams. Ask and you shall receive.

A dream is a direct reflection of the obstruction in the aura occurring simultaneously with the energy to heal this obstruction, each enfolded upon the other. As you align yourself with Divine Will and release the fear to know the Truth, you will discover the magnificent gift of the dream. For at each moment in the dream, there is the choice of truth or delusion. As you come to choose the Truth, so, too, will you come to transform yourself. Thus, dreams are both reflections and resolutions of the blocks in the aura, the Shadow, the darkness. It is helpful to keep a journal of your dreams and watch as they reflect your transformation. Your dreams may also assist in the transformation of others, as you open to the healing of yourself.

Since your dream is a reflection of your energy field, all elements of the dream, predominant or in the background are your creation. If you can experience identification with these elements and investigation their relationship, you may discover thought forms and fears not known before. Within the understanding of the relationship between these elements, or figures, in your dreams, you may come to know the healing of this relationship, which will be reflected in the healing within your auric field.

DREAM EXERCISE:

1. Remember a dream, preferably a recent dream. Imagine that

you are all the dream elements (e.g. symbols, persons, objects) in the dream, animate and inanimate.

2. As you explore the scene of the dream, say hello to each of these dream elements. Notice if they bring about any fear.

3. Select a dream element which holds the most fear.

4. Speak to this dream element, and then listen for a response. Speak back and forth to this object until there is understanding of the nature of this relationship.

5. Ask the dream element the following questions: What is the meaning of the fear? What is the fear asking of you? How can this fear be healed?

6. Allow the dream elements to speak freely with no judgment on your part.

7. Record their suggestions in your journal. Look for the manifestation of this healing in your daily life.

CHAPTER TWENTY SIX
RESPONSIBILITY AND COMMITMENT

Responsibility is the essence of work. The task to a responsible person is a duty, and duty is a most glorious endeavor if the attitude is one of generosity.

There may be a tendency to become entrapped in the obligatory containment and ignore the freedom present in each moment. Responsibility speaks to the fidelity to responsiveness and the joy of relationship to another and all the other's multi-faceted aspects.

This healing can occur within the moment or over a period of time. As consciousness expands, so, too, does darkness lighten. The space between the moments shrinks. Thus, rather than causing pain and fear, responsibility and commitment bring healing and the joy of knowing the All Mighty One.

As the soul's purpose unfolds and the uniqueness of the individual creation comes into form, there is an easing of the barriers placed between the self and other. As the struggle to become fully who you are opens to the Light, revealing you in all your glory, you will find the interest and desire in committing grows stronger. There is no fear of loss of Self or other, no fear of falling into the Void, never to be found again, no fear of separation. There emerges the knowledge of the Truth—that you and I and all are One. So there is pleasure in the joining of two fields in creative form.

In this commitment, karma and wounds from long ago are healed. The meeting in full love dissolves the pain, the irresolution of the past, replaced by alchemical creation. For in these meetings, potent creative energy is released that may be directed toward tasks on Earth. In opening to the desire to create together, there is enrichment offered to the Self, the other, and planet Earth.

In such meetings of souls there need not be pain instilled within another, for the moment of this commitment does not detract from the commitment to the other. All are full and real. The issue may be one of priorities of love. If the commitment is in love as man and woman to be faithful unto each other in the physical form, this becomes the highest commitment of the physical form. This commitment does not preclude love for and commitment to workmates, friends, work itself, to the soul's purpose, to a spiritual path, but it does remain the supreme commitment to the other as love partners.

The moment of commitment is of the highest creation. Trust that as you seek the wisdom held deep within your heart, there will be no pain, no limitation, no constraint within this joyful moment. The painful choices so envisioned will dissolve into the Light. You may have all and hurt no one as you trust in God's creation. Look always homeward to the heart. All the answers that you seek await your beckoning call.

Transferential projections obstruct commitment in that we do not see the partner in the full light of individuality but as the ultimate source of life-sustaining energy. The commitment becomes an addiction, an insatiable hunger for that which can never be. The other becomes the Great Mother, the Great Father, the anchor on Earth. In order to make a fully conscious commitment, we must dissolve these transferential glasses.

Acknowledging the Shadow is necessary for the addictive quality of commitment to dissipate. The fears of engulfment or abandonment inhibit communication and connection, further alienating the two that are attempting commitment.

The solution to this struggle is for each party to honor the Shadow and to acknowledge the fear and the desire to control the interaction. There is no right and wrong. This is not the issue. It is two beings seeking to love and be loved, enmeshed in the darkness of fear. By acknowledging the fear, the addictive hold loosens, and each can see the other as the humans that they are. True commitment cannot occur until the Shadow is acknowledged. For unless the Shadow reaches the Light, the commitment will be to an illusion and a fantasy.

Message:

Hold dear to your heart the precious gem to whom you feel responsibility, for there is honor in these relationships. Allow the freedom that you love to shine forth in the glory that is each individual creation. Know that responsibility is fostering the unfoldment of those you love.

Commitment is to hold the reins as your chariot races through the sky. It is to glow in the now of the glorious moment and to know the wholeness of your own Self. The wind, the storm, and the lightning flash through you. You are there in Truth.

Speak the Holy Name, and you shall know commitment. Heaven and Earth unite at the moment of commitment, becoming one in fearless love. Indeed, though fears of separation create the veil to hide the truth, at the very moment of commitment there is no fear. And is this not what you are seeking? A fearless state of All?

So burst forth in static ecstasy to know the Truth, and the freedom you seek will not be lost, but rather, found. It is the darkness of the fear of love that obscures the joy of committed being.

Commit to the moment and you will enter transmutative motion, for within this moment is the One and no other. As hearts open unto love, all will be fulfilled. The anticipated loss of opportunity within commitment evaporates within the fiery throngs of love. Rather than rushing on to the next moment in fear of loss of the veil of fear, surrender to the meeting and you shall know the One.

In responsibility there is the flow of destiny, as is also true within commitment. Hearts drawn as magnets to each other meet in the moment. This may occur in a momentary meeting or may be encapsulated within the form of friendship, workplace, marriage, spiritual path, neighborhood, family. Know that this meeting has meaning for you, whether it be of long or short duration. As you open unto love, so you will influence the love and commitment of the other.

The fear so often invested in responsibility and commitment is the fear of finding the Self you seek not to know, the darkness seeping through the cracks to remind you of its presence. For as intimacy intensifies, the mirror of the Self, as seen through the eyes of the other, shines more fully to your face.

You see yourself reflected back as the other responds to you. The dark corners of your being emerge to seek the Light. Thus, in responsibility and commitment is opportunity to heal the pain and fear held in these darkened spaces.

RELATIONSHIP AND COMMITMENT EXERCISE:

1. Sit quietly or lie down comfortably. Place your hand on your heart. Move inward, breathe, and relax.

2. Now move your right hand to your second chakra and your left hand to your heart. Begin to run energy from your right hand to your left hand, charging the second and fourth chakras.

3. Imagine the face of one with whom you are struggling about commitment. It may he a lover, a sibling, a child, a friend. Look into his or her eyes. What do you see? What do you feel?

4. Now let that face go and imagine your mother. Notice what you see and feel. Let her face go.

5. Imagine your father. What do you see in his eyes? What do you feel as you look into his eyes? Let his face go.

6. Let all the faces go and imagine yourself. Look into your eyes. What do you see and feel?

7. Now go back to each face in turn, and look into each person's eyes: (e.g. the one you are struggling with—your mother, your father), and tell one that you love them. Let them all go now.

8. Return to normal consciousness and write down what you would like to remember about the experience.

9. When you are finished with this exercise, put the writing away and return to it three days later. Additional insights may arise.

CHAPTER TWENTY SEVEN
THE WAVE OF 2012

We are immersed in an enormous wave of change as we move toward the year 2012 and beyond. Follow your inner guidance. Listen to your heart. It is a time most ripe for meeting your soul love and for moving into your soul's purpose here on Earth. We are blessed to be alive and completely supported by spiritual forces within, around, and outside the Earth's energy body.

As you read this book, if you can let the wisdom enter deeply into your heart, it will ease any potential pain of change and bring you to the glory that is yours.

Sit quietly and know that you and I are One.

ABOUT THE AUTHOR

As a healer, channel, and an acclaimed author of four books on metaphysics, Carolyn Cobelo has helped people all over the world connect to God and their highest selves. In the 1990's, she established the Akasha Institute of Spiritual Healing with locations in New York and Argentina. She has led pilgrimages to sacred temples and holy sites worldwide for over 25 years. She is also an artist, inventor of a spiritual board game (Avalon: Temple of Connection), and screenwriter. She is now creating documentary and feature films on metaphysical topics. She lives in Carmel, California where she is President of Akasha Entertaiment and the director of the Akasha Metaphysical Film Festival.

For more information on Carolyn Cobelo's other books, CDs, DVDs, and latest film projects, visit:

http://www.AkashaEntertainment.com

Also visit:
Twitter.com/Spiritual_Film
MetaphysicalFilm.com

Made in the USA
Lexington, KY
15 January 2012